BURLEIGH DODDS SCIENCE: INSTANT INSIGHTS

NUMBER 49

Improving water management in crop cultivation

Published by Burleigh Dodds Science Publishing Limited
82 High Street, Sawston, Cambridge CB22 3HJ, UK
www.bdspublishing.com

Burleigh Dodds Science Publishing, 1518 Walnut Street, Suite 900, Philadelphia, PA 19102-3406, USA

First published 2021 by Burleigh Dodds Science Publishing Limited
© Burleigh Dodds Science Publishing, 2022, except the following: Chapter 2 was prepared by a U.S. Department of Agriculture employee as part of their official duties and is therefore in the public domain. All rights reserved.

This book contains information obtained from authentic and highly regarded sources. Reprinted material is quoted with permission and sources are indicated. Reasonable efforts have been made to publish reliable data and information but the authors and the publisher cannot assume responsibility for the validity of all materials. Neither the authors nor the publisher, nor anyone else associated with this publication shall be liable for any loss, damage or liability directly or indirectly caused or alleged to be caused by this book.

No part of this publication may be reproduced, stored in a retrieval system or transmitted in any form or by any means electronic, mechanical, photocopying, recording or otherwise without the prior written permission of the publisher.

The consent of Burleigh Dodds Science Publishing Limited does not extend to copying for general distribution, for promotion, for creating new works, or for resale. Specific permission must be obtained in writing from Burleigh Dodds Science Publishing Limited for such copying.

Permissions may be sought directly from Burleigh Dodds Science Publishing at the above address. Alternatively, please email: info@bdspublishing.com or telephone (+44) (0) 1223 839365.

Trademark notice: Product or corporate names may be trademarks or registered trademarks and are used only for identification and explanation, without intent to infringe.

Notice
No responsibility is assumed by the publisher for any injury and/or damage to persons or property as a matter of product liability, negligence or otherwise, or from any use or operation of any methods, products, instructions or ideas contained in the material herein.

British Library Cataloguing in Publication Data
A catalogue record for this book is available from the British Library

ISBN 978-1-80146-286-0 (Print)
ISBN 978-1-80146-287-7 (ePub)

DOI 10.19103/9781801462877

Typeset by Deanta Global Publishing Services, Dublin, Ireland

Contents

1	Site-specific irrigation systems *Amir Hagverdi, University of California-Riverside, USA; and Brian G. Leib,* *University of Tennessee-Knoxville, USA*	1
	1 Introduction	1
	2 Field-level mapping of soil variability	3
	3 Delineation of irrigation management zones	6
	4 Quantifying the potential impact of variable rate irrigation	10
	5 Site-specific irrigation management	13
	6 Future trends and conclusion	18
	7 List of abbreviations	19
	8 Where to look for further information	20
	9 References	20
2	Deficit irrigation and site-specific irrigation scheduling techniques to minimize water use *Susan A. O'Shaughnessy, USDA-ARS, USA; and Manuel A. Andrade, Oak Ridge* *Institute for Science and Education, USA*	25
	1 Introduction	25
	2 DI strategies: overview	26
	3 DI strategies: approaches, risks and advantages	28
	4 SSIM: achieving precision irrigation	31
	5 Variable rate irrigation	33
	6 Integration of plant feedback sensor systems for site-specific VRI control	36
	7 Conclusions	39
	8 Where to look for further information	40
	9 Acknowledgements	40
	10 Disclaimer	40
	11 References	41
3	Improving water management in winter wheat *Q. Xue, J. Rudd, J. Bell, T. Marek and S. Liu, Texas A&M AgriLife Research and* *Extension Center at Amarillo, USA*	49
	1 Introduction	49
	2 Winter wheat yield	54
	3 Yield determination under water-limited conditions	55
	4 The role of measuring evapotranspiration (ET)	56
	5 Water-use efficiency	56
	6 Wheat yield, evapotranspiration (ET) and water-use efficiency (WUE) relationships	57
	7 Case studies	59
	8 Future trends and conclusion	66
	9 Where to look for further information	66
	10 References	67

4	Advances in irrigation techniques for rice cultivation	71
	D. S. Gaydon, CSIRO Agriculture, Australia	
	1 Introduction	71
	2 Water-saving measures	72
	3 Scale-dependency of water productivity and water savings	73
	4 Aerobic rice as a water-saving measure	75
	5 Alternate wetting and drying (AWD) as a water-saving measure	78
	6 Saturated soil culture (SSC) as a water-saving measure	81
	7 Case study: water-saving irrigation in southeast Australia	83
	8 Future trends and conclusion	85
	9 Where to look for further information	87
	10 References	87
5	Improving water management in sorghum cultivation	93
	Jourdan Bell, Texas A&M AgriLife Research and Extension Center, USA; Robert C. Schwartz, USDA-ARS Conservation and Production Research Laboratory, USA; Kevin McInnes, Texas A&M University, USA; Qingwu Xue and Dana Porter, Texas A&M AgriLife Research and Extension Center, USA	
	1 Introduction	93
	2 Dryland production	95
	3 Irrigation	96
	4 Deficit irrigation	99
	5 Soils and irrigation management	101
	6 Conclusion	103
	7 Where to look for further information	103
	8 References	103

Series list

Title	Series number
Sweetpotato	01
Fusarium in cereals	02
Vertical farming in horticulture	03
Nutraceuticals in fruit and vegetables	04
Climate change, insect pests and invasive species	05
Metabolic disorders in dairy cattle	06
Mastitis in dairy cattle	07
Heat stress in dairy cattle	08
African swine fever	09
Pesticide residues in agriculture	10
Fruit losses and waste	11
Improving crop nutrient use efficiency	12
Antibiotics in poultry production	13
Bone health in poultry	14
Feather-pecking in poultry	15
Environmental impact of livestock production	16
Pre- and probiotics in pig nutrition	17
Improving piglet welfare	18
Crop biofortification	19
Crop rotations	20
Cover crops	21
Plant growth-promoting rhizobacteria	22
Arbuscular mycorrhizal fungi	23
Nematode pests in agriculture	24
Drought-resistant crops	25
Advances in crop disease detection and decision support systems	26
Mycotoxin detection and control	27
Mite pests in agriculture	28
Supporting cereal production in sub-Saharan Africa	29
Lameness in dairy cattle	30
Infertility/reproductive disorders in dairy cattle	31
Antibiotics in pig production	32
Integrated crop–livestock systems	33
Genetic modification of crops	34

Developing forestry products	35
Reducing antibiotic use in dairy production	36
Improving crop weed management	37
Improving crop disease management	38
Crops as livestock feed	39
Decision support systems in agriculture	40
Fertiliser use in agriculture	41
Life cycle assessment (LCA) of crops	42
Pre- and probiotics in poultry production	43
Poultry housing systems	44
Ensuring animal welfare during transport and slaughter	45
Conservation tillage in agriculture	46
Tropical forestry	47
Soil health indicators	48
Improving water management in crop cultivation	49
Fungal diseases of fruit: Apples	50
Crops for use as bioenergy	51
Septoria tritici blotch in cereals	52

Chapter 1

Site-specific irrigation systems

Amir Hagverdi, University of California-Riverside, USA; and Brian G. Leib, University of Tennessee-Knoxville, USA

1 Introduction
2 Field-level mapping of soil variability
3 Delineation of irrigation management zones
4 Quantifying the potential impact of variable rate irrigation
5 Site-specific irrigation management
6 Future trends and conclusion
7 List of abbreviations
8 Where to look for further information
9 References

1 Introduction

1.1 Site-specific irrigation management

Conventional irrigation management is based on uniform application of water across a field which may result in both over- and under-irrigation. Over-irrigation negatively impacts soil aeration, increases disease occurrence and causes leaching/movement of mobile nutrients and sediments through the root zone into groundwater and off the field towards surface water bodies. Under-irrigation adversely affects yield quantity and quality, and leaves more applied nutrient in a field that could be lost due to winter rains. *Site-specific irrigation management relies on site-specific data to understand processes governing crop water requirement and/or soil water status on a spatio-temporal scale, and utilizes variable rate technology to optimize within-field allocation of applied irrigation water.* Note that the focus of site-specific irrigation is not only on the ability to spatially vary the application of water, but also on site-specific data acquisition and processing. The ultimate goal is to maximize the economic and/or environmental value of the water applied (Kranz et al., 2012) by (i) enhancing quantity and quality of production; (ii) reducing deep percolation of water and/or run-off from irrigated fields and in turn minimizing contamination of groundwater and surface water resources through nitrate leaching, transport of sediments, nutrient and agrochemical; and (iii) alleviating the large-scale impact of irrigation on water resources sustainability by optimizing the water footprint of irrigated agriculture. The basic assumption of site-specific irrigation is that crop water requirement may spatially vary due to many reasons that cause crop growth to be non-uniform across a field (Howell et al., 2012). Site-specific irrigation should emphasize the kinds of variability which can be managed by irrigation including changes in topography, soil texture/structure and drainage and salinity conditions. These are examples of parameters that impact infiltration, redistribution, storage and deep percolation of water, plant available water (PAW), leaching requirement and, consequently, characterize appropriate site-specific irrigation management strategies.

1.2 Variable rate irrigation development history

Most of the past work on site-specific variable rate irrigation (SS-VRI) has focused on pressurized mechanically moved overhead irrigation systems, mainly center pivot and lateral move. This is largely because of widespread adoption, adaptability to automation and the large area covered with a single lateral pipe (Sadler et al., 2005; Evans et al., 2013). Early developments included limited control of end guns, edge of the field stop and stop-in-slot control which allowed growers to stop the irrigation system after completion of irrigation events or at a specific position such as a field road. Later on, programmable control panels with speed control capability were developed. This feature made growers to be able to automatically apply different depths of water to a limited number of pie-shaped sectors (usually less than 10 sectors) underneath their center pivot irrigation systems, the earliest form of VRI. This method was often employed by growers when multiple crops were planted on different portions of the field (Kranz et al., 2012). More recent commercial products support greater speed control of irrigation machines resulting in more and smaller pie-shaped sectors (from <1° to 10°).

During the last two decades, a number of approaches have been developed to apply different depths of water along the moving lateral of overhead irrigation systems. This was a groundbreaking innovation in site-specific irrigation which led to the development of zone control technology for sprinklers that allowed delineation of irregularly shaped irrigation management zones (IMZs). In an early study by Roth and Gardner (1989) at the University of Arizona, a lateral move irrigation system was modified to apply different depths of water and nitrogen to a limited number of research field plots with the ultimate objective of developing water–nitrogen response surfaces. They used three spray lines each operated by separate solenoid valves while changing the travel speed of the machine. In the early 1990s, multiple site-specific irrigation research projects were initiated in the United States at Colorado (Fraisse et al., 1995a,b), University of Idaho (King et al., 1995), South Carolina (Omary et al., 1997) and Washington State University (Sadler et al., 2005).

In Idaho, King et al. (1995) studied site-specific application of water and chemicals using both lateral move and center pivot irrigation systems which consisted of a spatially referenced mapping system. In another study, King and Kincaid (2004) designed a variable flow sprinkler by changing the nozzle area using a mechanically activated pin. The main concern about this design is that changes in flow rate alter wetting patterns and water droplet size distribution of sprinklers, which in turn may create application uniformity issues (Kranz et al., 2012). In Colorado, Fraisse et al. (1995b) used the concept of pulse irrigation defined by Karmeli and Peri (1974) as a series of on (operating phase) and off (non-operative phase) irrigation cycles on a lateral move irrigation system to apply different amounts of water along the system. The main objective of Colorado State University's experiment was to perform small-plot research on conjunctive management of irrigation water and chemicals using a lateral move system. However, they also mentioned potential applications of VRI to address spatial differences such as soil type and topography. In South Carolina, VRI was designed and constructed in cooperation with the University of Georgia on a center pivot irrigation system using a multiple manifold technique (Camp et al., 1998). In Washington, researchers also used multiple manifolds with different nozzle sizes to develop a variable rate center pivot irrigation system. Further developments in the Washington State University focused on electronic controllers for the activation of solenoid valves to control a bank of sprinklers on a center pivot irrigation system (Sadler et al., 2005). Currently, commercial VRI systems usually use a combination of speed control and pulse irrigation. More recent efforts such as studies by Han et al. (2009), O'Shaughnessy et al. (2015) and Chávez et al. (2010a,b) focused on integration of VRI technology with site-specific data acquisition and processing systems using wireless technologies, geographical information system and global navigation satellite system (GNSS) applications, Internet and soil/plant monitoring systems.

The focus of this chapter is on SS-VRI using center pivot and lateral move irrigation systems which are mainly used for row-crop irrigation. However, basic concepts discussed throughout the chapter on site-specific data acquisition and mining approaches, soil mapping, zone delineation, site-specific yield estimation and irrigation scheduling should be applicable to other existing forms of site-specific irrigation technologies and/or new forms that may become available in the future for fixed sprinkler, drip and surface irrigation systems, as well as potential applications in nursery and urban irrigation settings.

2 Field-level mapping of soil variability

Mapping field-level soil heterogeneity is an essential prerequisite for almost all precision agriculture (PA) applications. Field-level spatial information of soil hydraulic attributes [mainly field capacity (FC) and permanent wilting point (PWP), which are used to calculate PAW] is required for SS-VRI management. However, direct measurement of these characteristics is challenging due to the time-consuming and labour-intensive nature of *in situ* and laboratory methods. The most widely used solution to this problem is to develop proxies of soil hydraulic attributes by collecting easily measured soil characteristics such as soil textural information (i.e. per cent of sand, silt and clay), soil bulk density (BD) and organic matter (OM) content that are typically well-correlated with soil hydraulic properties to produce pedotransfer functions (PTFs, Bouma 1989). There are many PTFs in the literature that provide non-spatial point estimation of soil water retention information. However, the literature is limited with regard to the production of high-resolution maps of soil hydraulic attributes.

The PA community predominantly focuses on both on-the-go sensors and remote sensing (RS) to map soil variability, with soil apparent electrical conductivity (ECa) being the most widely used proxy. ECa is a function of the electrical conductivity of porous media solution, the soil porosity and the cementation exponent, that is, Archie's law (Archie, 1942). When soil salinity is not a major factor, ECa may be a useful proxy of soil physical and hydraulic attributes (Sudduth et al., 2005) including depth to sand layer (Duncan, 2012), clay percentage under non-saline conditions (Saey et al., 2009) and soil texture and FC (Abdu et al., 2008). Some site-specific irrigation studies that focused on ECa to characterize soil variability have been conducted by Hedley and Yule (2009a,b), Pan et al. (2013) and Haghverdi et al. (2015a). These endeavours introduced and evaluated novel approaches to map spatial heterogeneity of soil hydraulic properties at high resolution using ECa as the main proxy.

2.1 Case study: high-resolution prediction of plant available water

This case study draws from previous work conducted and published by Haghverdi et al. (2015a) on a 73-ha field located in West Tennessee close to the Mississippi River. The soil survey indicated that the field contains Mississippi River terrace alluvial deposits that have produced Robinsonville loam, fine sandy loam, Commerce silty clay loam and Crevasse sandy loam soils. On 20 March 2014, ECa was measured at 4700 points at shallow (0–30 cm) and deep (0–90 cm) depths across the study area using a Veris 3100 machine (Veris Technologies, Salina, KS, USA) after some rainfall events when soil was assumed to be close to FC. Next, field soil sampling was conducted. On 21–22 March 2014, a truck-mounted hydraulic probe was used to sample 100 undisturbed sites at 0–100 cm depth where each soil sample was divided into four segments. Hereafter, the word 'layer' is used to distinguish among segments rather than real soil horizons. The default depth of subsamples was 25 cm, though adjustments that accounted for soil horizon transitions were made. Soil textural information (i.e. per cent of sand, silt and clay), BD and gravimetric water content were measured in the laboratory.

Ten spatial modelling scenarios that used kriging (KG), co-kriging (CKG), regression kriging (RKG), geographically weighted regression (GWR) and artificial neural network (ANN) were examined (Fig. 1). The objectives were to convert measured soil properties at point locations to continuous maps of water content at FC and PWP, and to evaluate the efficiency of the ECa as a proximal attribute in this process. FC and PWP are widely used

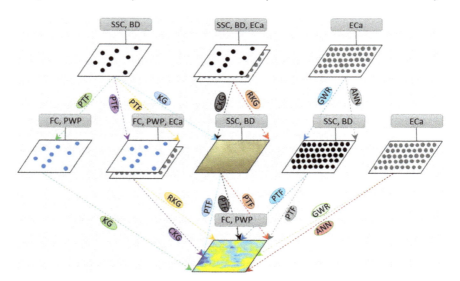

Figure 1 Ten spatial modelling scenarios using pseudo-continuous pedotransfer function (PTF, Haghverdi et al., 2012, 2014), kriging (KG), co-kriging (CKG), regression kriging (RKG), geographically weighted regression (GWR) and ANN. The inputs were per cent of sand, silt and clay (SSC), bulk density (BD, Mg m^{-3}) and soil apparent electrical conductivity (ECa, mS m^{-1}). The outputs were soil water content maps at field capacity (FC, cm^3 cm^{-3}) and permanent wilting point (PWP, cm^3 cm^{-3}) (adapted from Haghverdi et al. (2015a)).

thresholds for irrigation management. FC is a qualitative parameter that is a practical and understandable indicator of soil water holding capacity (Romano and Santini, 2002). Water retentions at −10 and −33 kPa could be considered as FC for coarse- and fine-textured soils, respectively (Rivers and Shipp, 1972). The water content at −1500 kPa is usually chosen as the PWP for all soil textures. For this study, it was assumed that at FC, matric potential was uniform across the study area (−10 kPa). PAW is obtained as the difference between water content at FC and PWP within the root zone.

2.2 Summary of findings

2.2.1 Performance of the modelling scenarios

The accuracy of soil water content maps improved when ECa was incorporated in the workflow models. Two methods showed higher performance than others: (i) ECa was required as input to establish a GWR model on discrete sampling locations which was subsequently used to generate maps of FC and PWP, and (ii) PTF was applied to predict water retention at FC and PWP on discrete sampling locations, and subsequently those predictions together with ECa data were used to generate maps of FC and PWP using RKG or CKG as the interpolation approach.

2.2.2 Application of ECa

Proximal and easily collected data such as ECa are affected by a variety of soil- and/or crop-related properties, hence should be applied and interpreted very cautiously. Unfortunately, complexity of proximal sensor measurements and in particular ECa mapping are sometimes overlooked in research studies (Corwin and Lesch, 2010) by the PA community and to a greater extent in practical applications by growers and PA professionals. The key site-specific question that should always be precisely answered is what underlying processes and soil–plant attributes affect ECa measurements. In the field of study, intensive soil sampling revealed that the spatial distribution of soil texture was the main factor governing soil water retention at all layers, and ECa measurements showed a high correlation with soil textural properties and soil water status at the time of sampling. Some studies showed that it is better to conduct ECa mapping under wet conditions because the relationship between ECa and texture is more pronounced near FC (Auerswald et al., 2001; Earl et al., 2003). Useful guidelines and insights on the application of ECa in PA have been provided by Corwin and Lesch (2003) and by Corwin and Lesch (2010).

Since sampling is time-consuming and expensive, it is desirable to minimize sampling density. However, one should note that the sampling density and scheme affect the error associated with the spatial prediction process (Herbst et al., 2006). Sampling at half of the variogram range (if there are existing variograms of the soil/crop properties) is suggested as a 'rule of thumb' to collect enough samples to make reliable prediction for PA (Kerry et al., 2010). Debaene et al. (2014) showed that 1.5 samples ha^{-1} is adequate to predict soil OM content and texture. Iqbal et al. (2005) recommended sampling intervals of <100 m in order to detect boundaries of soil hydraulic properties in an alluvial floodplain soil in the region of the Mississippi Delta. Taking 1 sample ha^{-1} is usually considered the greatest sampling density that a farmer can afford (Kerry et al., 2010). However, an additional concern of site-specific irrigation management is to collect deep enough samples to sufficiently cover the effective water uptake zone by crop roots.

Incorporation of ECa or similar easily collected dense proxies in the spatial mapping process is suggested either as a guide for directed subsequent sampling or as an input for spatial modelling. This alleviates the need for more difficult data collection (such as soil texture and BD) through sampling, which in turn reduces the associated cost of the mapping process while maintaining an adequate accuracy level.

3 Delineation of irrigation management zones

A management zone (MZ) is generally defined as a subregion of a field that is relatively homogeneous with respect to soil–landscape attributes. However, the phrase 'management zones' remains uncertain unless additional information is included to clarify the goal in subdividing the field (Kitchen et al., 2005). In PA, the focus of available studies has been mainly on delineation of yield-based productivity zones and MZs for variable rate application of fertilizer (precision nutrient management). In contrast, there are only a handful of studies on IMZ delineation (e.g. Haghverdi et al., 2015b, 2016; Mouazen et al., 2014; Boluwade et al., 2015; Fortes et al., 2015), while there is a critical need to dynamically develop approaches for IMZ delineation in an accurate and inexpensive manner (Evans et al., 2013). In order to accomplish this objective, appropriate attributes and algorithms for IMZ delineation need to be identified along with some criteria to define optimum number of IMZs for a given field.

Delineating zones for SS-VRI management is challenging. The ultimate goal should be to identify and group together those areas within a field that need identical irrigation treatment. A variety of soil-related attributes such as topography, restricted subsurface soil layers, soil type and soil salinity may be important for IMZ delineation. In practice, mainly easily collected inexpensive non-invasive proximal data collected using RS techniques or on-the-go sensors (e.g. yield maps, topography, satellite photographs, canopy images and ECa) that are representative and spatially correlated to these soil properties are used for zoning. As stated earlier in this chapter, the main drawback of proximal sensor measurements is that they are complex and affected by multiple soil-crop properties (Corwin and Lesch, 2010). Therefore, they should show similar spatial distributions to main factors of concern in irrigation decisions, in order to be considered as an appropriate input to delineate IMZs.

Applying unsupervised clustering techniques and zoning via user-defined thresholds are the two main methods to delineate MZs. Clustering techniques such as *k*-means and fuzzy *k*-means use the inherent structure and distribution of the data in order to group similar observations together into distinct classes and have been widely used to identify MZs (Córdoba et al., 2013). These methods usually produce separated irregular (freeform) shape zones across a field. Integer linear programming (ILP) can be used to make rectangular and pie-shaped zones (Cid-Garcia et al., 2013; Haghverdi et al., 2015b).

3.1 Case study: delineating management zones for variable rate irrigation on a no-tillage cotton field in West Tennessee

This case study was conducted and published by Haghverdi et al. (2015b) on the aforementioned field in western Tennessee which contained two center pivot irrigation

Site-specific irrigation systems

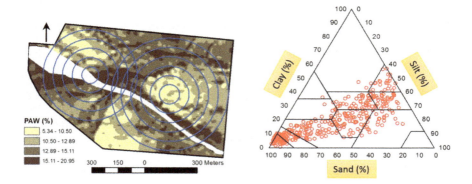

Figure 2 Left panel: Map of plant available water (i.e. PAW: FC-PWP) produced by geographically weighted regression (GWR) model for the entire sampling depth (1 m) [adapted from Haghverdi et al. (2015b)]. Right panel: The distribution of soil texture in samples from the field of study (adapted from Haghverdi et al. (2015a)).

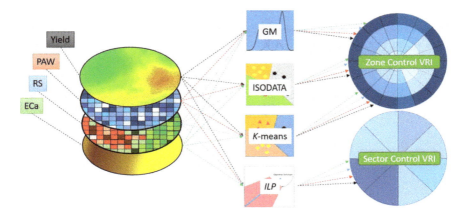

Figure 3 Multiple strategies to delineate IMZs for zone control and sector control variable rate irrigation (VRI) center pivot machines. The clustering approaches are Gaussian mixture (GM) model, ISODATA-maximum likelihood (ISODATA), k-means and integer linear programming (ILP). The inputs attributes are yield, plant available water (PAW, estimated as explained in part 2), space images from Landsat 7 and 8 satellites (RS) and soil apparent electrical conductivity (ECa) (adapted from Haghverdi et al. (2015b)).

systems that were used for supplemental irrigation (Fig. 2). The objectives were to evaluate multiple input attributes and algorithms for IMZ delineation, and then identify the optimum number of zones based on a particular attribute. Three unsupervised clustering techniques (i.e. k-means, ISODATA-maximum likelihood and Gaussian mixture model) with different combinations of input soil physical and proximal attributes (i.e. PAW, ECa, yield data and Landsat 7 and 8 satellite images) were examined (Fig.

3). Most center pivot irrigators prefer to have less than 10 MZs in a field (Evans et al., 2013). Therefore, clustering procedures were repeated to delineate from 2 to 10 IMZs. In addition, an optimization problem was formulated using the ILP technique to find the optimum number and positions of pie-shaped zones underneath the two center pivots. Pie-shaped zones are appropriate for center pivot machines with speed control capabilities (sector control VRI) where speeding the system up and slowing the system down provide less and more time for irrigation, respectively. Machines with sector control capabilities only adjust travel speed and do not provide adequate flexibility to fully manage within-field heterogeneity because sprinklers apply the same irrigation depth along the entire lateral, lacking the capability to delineate irregularly shaped zones (Kranz et al., 2012). However, this is a relatively easy-to-operate and affordable VRI strategy for those growers who possess center pivot irrigation systems because most of the available center pivots have control panels with a speed control module. The knowledge and full use of this capability provide a great opportunity to get growers acquainted with the VRI concept and bridge the knowledge and experience gap between current conventional uniform irrigation practice and cutting-edge commercially available zone control technology.

The per cent of total variance of PAW (following dividing the field into a given number of IMZs) was used as an evaluation measure to identify the optimum number of IMZs. This criterion indicates how much variability is explained as the number of IMZs increases, and was calculated following Fraisse et al. (2001) and Brock et al. (2005). First, the PAW variance was determined for each zone and weighed considering the zone area (Equation 1), then summed up to obtain the overall within-zone variance of PAW (Equation 2), and finally converted to the per cent of total variance (Equation 3):

$$VAR_Z = \frac{1}{n_Z} \sum_{i=1}^{n_Z} (PAW_i - PAW_Z)^2 \tag{1}$$

$$VAR_T = \sum_{i=1}^{m} VAR_i \times \frac{n_Z}{n_T} \tag{2}$$

$$\% \, of \, total \, VAR = 100 \times \left(VAR_T \big/ VAR_{WF} \right) \tag{3}$$

VAR_Z: weighted variance for zone Z
PAW_i: plant available water for cell i (25 m² in this case study)
PAW_Z: mean plant available water in zone Z
n_Z: number of cells in zone Z
VAR_T: summed and weighted within-zone variances
n_T: total number of cells across field
m: number of zones
VAR_{WF}: variance calculated for all cells across the field (represents conventional uniform irrigation with only one zone)

3.2 Summary of findings

3.2.1 Zoning algorithms

The unsupervised clustering algorithms provided similar results and good agreement among delineated zones using the clustering techniques (i.e. *k*-means, ISODATA-maximum likelihood and Gaussian mixture model) observed. The findings indicate that for a field with high spatial structure, all well-known clustering methods recognize similar patterns for IMZs.

3.2.2 Input attributes

Soil ECa followed by the satellite images showed high potential for IMZ delineation. Successful applications of ECa for zone delineation were also reported by Moral et al. (2010) and Kitchen et al. (2005). The spatial pattern of captured data by space images was affected by soil water status and crop phenology pre- and in-season, respectively. Not all space images were appropriate for IMZ delineation; therefore, it is important to screen and find reliable images. In practice, farmer's knowledge on field conditions could be extremely useful to choose the most appropriate data set and this knowledge should be incorporated in the zone delineation process. Yield data indicated less potential for zoning and showed some inconsistency mainly because different factors including physical and chemical properties of soil, irrigation regimes, pests and diseases and climate affect yield in a complex manner. Averaging yield maps across years could help find a more stable scheme of yield distribution, yet productivity zones may not necessarily match the irrigation zones.

3.2.3 Optimum number of zones

The results of this case study showed that having more than 4 to 5 zones only slightly reduced the per cent of total variance (Fig. 4), hence did not improve the ability to explain soil heterogeneity, a trend also reported by Zhang et al. (2010). Note that IMZ maps are not identical to irrigation prescription maps. Site-specific data on soil/crop water status are needed to convert MZ maps to irrigation prescription maps. It is expected that management cost (i.e. cost of instrumentation, measurements and data interpretation) increases as the number of zones increases. Therefore, it is important to find the optimum number of IMZs for each field. Having too many IMZs means making too many irrigation decisions, something that can easily overwhelm growers.

3.2.4 Dynamic in-season zoning

If in-depth vertical soil variation is significant, the spatial arrangement of zones may vary throughout the growing season considering the expansion of the effective root zone. In this case, multiple zoning schemes may be needed within a cropping season (Haghverdi et al., 2015b). In practice, the number and spatial arrangement of IMZs may be influenced by a variety of parameters including available water for irrigation, crop response to over/under-irrigation and temporal variability in seasonal climatic parameters. Continuous monitoring of soil water status and/or plant stress throughout the season helps in adjusting spatial arrangement of IMZs, if needed.

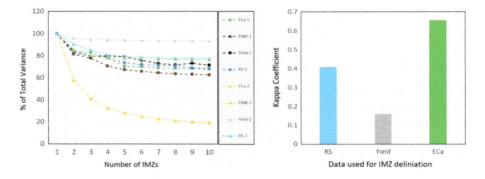

Figure 4 Performance of different input attributes and number of irrigation management zones (IMZs) averaged for the two center pivot systems in the study site. Left panel shows per cent of total variance (Equation 3) and input attributes were ECa (soil apparent electrical conductivity), PAW (plant available water, estimated as explained in part 2), yield (cotton lint yield in 2013 and 2014) and RS (reflectance values from panchromatic band taken by Landsat 7 and 8 satellites). Letter Z indicates zoning was done for zone control VRI scenario using k-means approach, and letter S shows zoning was conducted using integer linear programming (ILP) for sector control VRI strategy. Right panel depicts kappa coefficient (Cohen, 1960) values for the zones delineated by ILP. The kappa coefficient shows inter-classification agreement of different zoning strategies where a positive coefficient indicates that agreement exceeds chance and the magnitude of the coefficient reveals the agreement strength (adapted from Haghverdi et al. (2015b)).

4 Quantifying the potential impact of variable rate irrigation

It is important to understand how site-specific irrigation technologies and management strategies affect crop production for a given field. Currently, it is not easy to quantify the potential production benefits of SS-VRI irrigation systems over conventional uniform irrigation strategies, which contributes to growers' present uncertainty in making investment in VRI systems. There are only a few studies in the literature documenting the impact of SS-VRI on irrigation water use efficiency (WUE) and crop yield, and the reported results are not consistent among studies. Bronson et al. (2006) evaluated cotton lint yield response to applied irrigation rate for different landscape positions (i.e. bottom slope vs side slopes) and reported no difference between the two, thereby suggesting that VRI would not be beneficial at the experimental site. King et al. (2006) studied the impact of SS-VRI systems on potato yield in comparison with conventional uniform irrigation. Soil moisture sensors were used to adjust water application based on estimated evapotranspiration (ET) at each MZ, while conventional uniform irrigation was to base irrigation depth on the average irrigation requirement for the experimental units. They reported 4% greater yield under SS-VRI scenario with an increase in income of about half the cost of the technology. Some other studies (e.g. Vories et al., 2016; Sui et al., 2015) have used VRI systems, yet mainly to apply different amounts of water to irrigation research plots without investigating the impact of VRI versus uniform irrigation to address field-level soil variability.

4.1 Case study: evaluating rainfed, uniform and variable rate irrigation strategies

This case study was conducted and published by Haghverdi et al. (2016) on the aforementioned field located in West Tennessee with the mean monthly growing season temperature and precipitation equal to 21 °C and 97 mm month^{-1} from May to November, respectively. A two-year cotton irrigation experiment (2013–14) was implemented. The main objective was to estimate the potential impact of SS-VRI systems on cotton production and irrigation water use. Therefore, a variety of site-specific data were collected and used to design and evaluate some site-specific water production functions (SS-WPFs). The best SS-WPF was selected to generate yield maps for four irrigation and zoning strategies: (i) rainfed, (ii) conventional uniform supplemental irrigation, (iii) sector control VRI and (iv) a hypothetical zone control VRI capable of irrigating each cell (25 m^2) at an optimum level. For sector control VRI, the aforementioned ILP approach was adopted to find the optimum arrangement and number of IMZs, while this time the optimization objective was to maximize cotton lint yield. For preprocessing the data (i.e. four categories: soil, irrigation, fertilizer and yield), principal component analysis and k-means were used and, for deriving WPFs, multiple linear regression, ANN and k-nearest neighbour (k-NN) techniques were employed (Fig. 5).

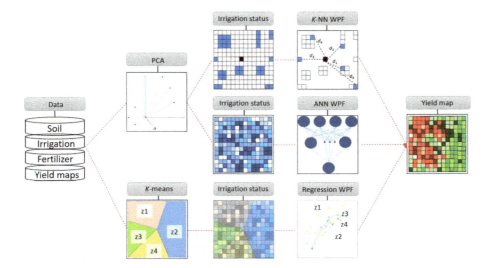

Figure 5 The three modelling scenarios used to develop site-specific water production functions (SS-WPFs) to predict cotton yields for the two center pivots in the study site. The input/output attributes in soil category: (soil apparent electrical conductivity, soil textural information, soil bulk density, soil volumetric water content at the time of sampling, at field capacity and at permanent wilting point); in crop category: (cotton lint yield in 2013 and 2014); in irrigation category (seasonal applied irrigation in 2013 and 2014); and in fertilizer category (maps of applied phosphorus and potassium by the grower). The statistical approaches used for preprocessing of data and subsequently development of WPFs were PCA: principal component analysis, k-means, k-NN: k-nearest neighbour, ANN: artificial neural networks and regression (adapted from Haghverdi et al. (2016)).

4.2 Summary of findings

4.2.1 Performance of the modelling approaches

K-NN WPF showed the highest performance (root-mean-square error equal to 0.131 Mg ha^{-1} and 0.194 Mg ha^{-1} in 2013 and 2014, respectively). K-NN is a non-parametric technique and works as a smart screening engine to select only a few relevant data points for prediction. This approach helps k-NN WPF to take advantage of the high number of yield data points to distinguish patterns without fitting coefficients. One should note that k-NN WPF introduced in this study provides discrete predictions of yield over a range of applied irrigation, hence providing only reliable estimations in the neighbourhood of measured applied irrigation depths.

Site-specific WPFs are examples of data-driven models that could be developed using farming data generated nowadays by PA. These models could be used as after-the-fact tools to evaluate different irrigation and zoning strategies and to further enhance site-specific irrigation management. When water is limited for full irrigation, SS-WPFs can be used to find the optimum deficit irrigation strategy to allocate water among IMZs. In addition, SS-WPFs are useful to identify the best irrigation strategy (which soil to follow) for zones with a mixture of soils which is likely to happen in pie-shaped zones delineated for sector control VRI systems. Note that the performance of data-driven models such as SS-WPFs may differ and should be tested for different soils, crops and climatic conditions. To derive robust SS-WPFs, all major attributes affecting crop response to irrigation should be incorporated in the modelling process.

Readers should consider that there are fundamental differences in characteristics and applications of the traditional crop WPFs and proposed SS-WPFs in this chapter. Classical WPFs are typically established using data collected from research plots through multi-year experiments in controlled conditions. The classical WPFs are developed for reuse in similar places and/or for the same place in the future. In contrast, the SS-WPFs are designed for site-specific use to convert site-specific farming data to useful information, an opportunity that has only become possible by recent advances and the spread of precision farming equipment which has led to collection of numerous site-specific data by growers.

4.2.2 Impact of site-specific irrigation strategies on cotton lint yield

The average estimated cotton lint yields in 2013 and 2014 by k-NN SS-WPF for the two center pivots in the study site were: rainfed (100%, 0.94 Mg ha^{-1}), uniform supplemental irrigation (139%, 1.3 Mg ha^{-1}), sector control VRI with 10 IMZs (144%, 1.35 Mg ha^{-1}) and the hypothetical high-resolution zone control VRI (155%, 1.45 Mg ha^{-1}). High in-season rainfall is typical in West Tennessee, yet some dry periods occur when supplemental irrigation is crucial to fulfilling ET demand. Over the course of this experiment, the applied supplemental irrigation was less than 20% of precipitation. The impact of site-specific irrigation strategies on crop production strategies is expected to be more pronounced in drier years. Preseason wet soil condition and colder temperatures forced the farmer to postpone planting cotton in 2013, which impacted yield pattern (Fig. 6) and crop response to irrigation mainly due to the inability of the cotton to accumulate adequate heat units.

The main objective in crop-based zoning using WPFs is to maximize production, while in soil-based zoning (explained in Section 3), the main objective is to maximize explained soil

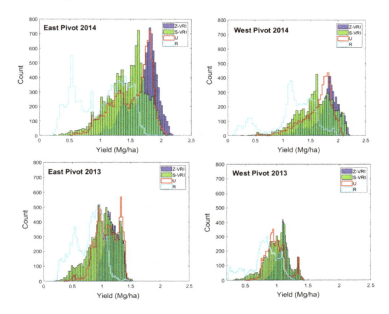

Figure 6 Histogram of cotton lint yield estimated using site-specific WPF (developed by k-nearest neighbour) for the center pivots in the field of study. Yield data were estimated in 2013 and 2014 cropping seasons for the four different water management strategies: R (rainfed), U (conventional uniform), S-VRI (sector control VRI with 4 IMZs) and Z-VRI (a hypothetical zone control VRI with capability to irrigate each 25 m^2 cell at the optimum level) (adapted from Haghverdi et al. (2016)).

variability. The crop-based zoning using WPFs is influenced by annual climatic conditions and farmer management practices. A long-term farming data record that captures different climatic and irrigation management scenarios is expected to lead to more stable IMZs. In practice, it is suggested to start with soil-based zoning and then analyse crop response to irrigation depths across zones using farming data and site-specific models (such as k-NN WPF) to improve arrangement and number of IMZs in an iterative process.

5 Site-specific irrigation management

5.1 Prescription maps for variable rate irrigation

The two major approaches for variable rate application of inputs are (i) map-based: site-specific data are collected and converted to prescription maps prior to implementation and (ii) sensor-based: real-time sensor data are collected and converted to application rate via a pre-developed application algorithm. For SS-VRI systems, ideally a combination of the two approaches is needed to first delineate IMZs, and then throughout the cropping season, create an irrigation prescription map before each irrigation event using real-time site-specific data collected via sensors. The three main methods that have been used/ proposed for irrigation scheduling using SS-VRI systems are (i) soil moisture-based (Liang et al., 2016; King et al., 2006; Kim et al., 2009; O'Shaughnessy et al., 2015), (ii) ET-based (Hedley and Yule 2009b) and (iii) plant feedback-based (O'Shaughnessy et al., 2015) approaches.

Liang et al. (2016) deployed a network of soil moisture sensors to measure soil matric potential across IMZs and subsequently converted the data to equivalent volumetric water content information using the soil water retention equation of van Genuchten (1980) to create irrigation prescription maps. In another study, O'Shaughnessy et al. (2015) used a field-calibrated neutron probe to measure soil water content on a weekly basis in order to calculate the replenishment of crop water use to FC in the top 1.5 m of soil. King et al. (2006) measured soil water contents using TDR soil moisture sensors which were subsequently used to adjust water application based on estimated ET at each MZ. The main practical question for the soil moisture-based approach is the number, location and depths of soil moisture sensors which should be installed per IMZ. Hedley and Yule (2009b) conducted a simulation study to run a water balance model based on ET, to irrigate each zone when a critical soil moisture deficit is reached. O'Shaughnessy et al. (2015) introduced a plant feedback method which uses meteorological data and a wireless network of infrared thermometers (mounted on the pivot lateral and placed in the well-watered spots throughout the field). In their case study, the pivot scanned the field on even days and irrigated using an automatically created prescription map on odd days.

5.2 Uniformity and spatial resolution of variable rate systems

It is necessary to evaluate application uniformity of SS-VRI systems and assess their reliability to apply different depths of water and change the application rate as prescribed. Table 1 summarizes the reported results from several case studies by Sui and Fisher (2015), Han et al. (2009), O'Shaughnessy et al. (2013), Dukes and Perry (2006) and Yari et al. (2017) on application uniformity of various-size center pivot or lateral move systems equipped with different types of sprinkler packages. The drop hoses were spaced from 1.5 m to 3 m apart and pressure regulators were rated at 41–138 kPa. Some systems were equipped with commercially available VRI technologies, while the others were outfitted with prototype technologies developed by researchers. Application rates varied from 11% to 100% (100% means constant application with no cycling) with different equivalent water depths. The collectors were put in a variety of grid, arc-wise and transect patterns. Overall, coefficient of uniformity (CU) and low quarter distribution uniformity (DUlq) averaged about 88% (min: 73% and max: 95%) and 78% (min: 73% and max: 95%) between studies, respectively.

Han et al. (2009) and Sui and Fisher (2015) reported some reductions in CU for lower application/cycling rates. However, the overall impact of application/cycling rates on CU was not noticeable between studies. Reported results indicate that SS-VRI systems adequately applied the desired irrigation depths in most cases. Sui and Fisher (2015) reported approximately on average 13% difference between actual and desired application amounts and noticed higher error for lower application depths. O'Shaughnessy et al. (2013) observed less than 3 mm error (root-mean-square error) between actual and desired application depths. They reported that some variations in depths of applied water are expected due to wind direction and speed. Dukes and Perry (2006) observed a noticeable influence by nozzle type on the uniformity and attributed that to the distinct spray pattern of each sprinkler type.

Site-specific irrigation systems

Table 1 Summary of some reported results on uniformity and performance of SS-VRI systems

			System					Sprinkler						Test		
Ref.	Type	Make[a]	VRI	Spans	Length (m)	Press[f] (kPa)	Flow (m3/min)	Type[e]	Height (m)	Spacing (m)	Press (kPa)	WET (m)	Wind (m/s)	application	Collectors	CU	DUlq
Dukes and Perry (2006)	CP	Reinke	developed by researchers and Computronics Corp. Ltd.	4	186	262	0.9	Nelson R3000 rotator sprinkler nozzles	1.3	3.05	67		<3.8	20%, 50%, 80%, 100% at two speed rates 7% and 11% (100% @ 11% speed was 9mm)	2 outer spans- 180 collectors 6 lines (3 reps) Spacing: 3m by 15 m	89–95%	84–95%
Dukes and Perry (2006)	LM	Valley	developed by researchers	6	375	206[f]	3.8	Senninger LDN (fixed plate)	1.5	2.3	138		<5	25%, 50%, 75%, 100% at two speed rates 11% and 16% (100% @ 11% speed was 20–21mm)	2 centre-most spans- 108 collectors 6 lines (3 reps) Spacing: 3 m by 7.6m	73–83%	60–74%
								Senninger i-WOB (moving plate)	1.8	3	138		<5			84–91%	73–82%
Sui and Fisher (2015)	CP	Valley	Valley VRI	4	233	234*/290	1.33	Fixed-pad Senninger LDN-UP3	1.8	2.74	103	7–13.4	3.58	30%, 50%, 70%, 100% (100% was 25.4 mm)	4 spans 210 collectors 2 lines Spacing: 2.44m with 12° between lines	78–89%	
Yari et al. (2017)	CP	Valley	Valley VRI	5	294	240		LESA, Nelson R3000 rotator	1.5	2.29	120		2.4	8.1mm, 14.2mm, 20.3mm	spans 2–5 Spacing: 3 m with 4° between lines	*92–94%	
O'Shaughnessy et al. (2013)	CP	Valley	Valley VRI	3	131	172		Fixed plate Senninger LDN	1.5	1.5	41	6–9.0	4.1–9.5	30%, 50%, 70%, 80%, 100% (two travel speeds)	spans 2–3 transect, arc-wise, and grid patterns	86–90%	76–83%
	CP	Valley	Valley VRI	6	260	172		Fixed plate Senninger LDN	1.5	1.5	41	6–9.2	2–7.9	30%, 50%,70%, 80%, 100%	Spans 5–6 transect, arc-wise, and grid patterns	86–90%	76–83%

(Continued)

Table 1 (Continued)

Ref.	System							Sprinkler						Test		
	Type	Make[a]	VRI	Spans	Length (m)	Press[f] (kPa)	Flow (m3/min)	Type[€]	Height (m)	Spacing (m)	Press (kPa)	WET (m)	Wind (m/s)	application	Collectors	CU	DUlq
King et al. (2005)	LM		developed by researchers	3	100				1.5	3	138		<2.5	36%, 52%, 68%, 84%, 100%	multiple grid patterns	86–95%	
Han et al. (2009)	LM	Reinke	developed by researchers	2	76	172		Nelson 36 quick-change spinner nozzles	2	3	140			100%, 75%, 50%, 25% (100 % was 25 mm)	68 collectors spacing: 0.75m × 1.5 m grid	80–95%	

[f] Dukes and Perry (2006): inlet pressure was in non-variable-rate irrigation mode and at the pivot point. Sui and Fisher (2015): inlet pressure at the pivot point was 234 kPa and 290 kPa for the constant application rate and for variable application rate test, respectively.
[€] LESA: low elevation spray application; LDN: low drift nozzle; CP: center pivot; LM: lateral move; WET: Sprinkler wetted radius; Nelson Irrigation Corporation (Walla Walla, Washington); Senninger by Hunter Industries (Clermont, Florida).
[¥] Yari et al. (2017): reported results are for the experimental set-up on 4 September 2014.
[a] Reinke Manufacturing Company, Inc.: (Deshler, Nebraska, USA); Valley Irrigation by Valmont Industries, Inc.: (Omaha, Nebraska, USA).

Site-specific irrigation systems

There is a transition area between adjacent zones in which application depth changes gradually which causes edge effect between adjacent IMZs both along the irrigation system and in travel direction (Kranz et al., 2012). That is because, for a typical moving overhead irrigation system, there is an overlap between wetting areas of sprinklers such that each spot on the ground is irrigated by multiple sprinklers. The edge effect reduces application uniformity when different depths of water are applied in each zone (Sui and Fisher, 2015; O'Shaughnessy and Evett, 2015). The width and length of edge effect mainly depend on wetted diameter of sprinklers and overlap pattern, yet other factors including wind (speed and direction) and GNSS accuracy may change the location and shape of the area of influence. Chávez et al. (2010b) reported an average positioning error of 2.5 ± 1 m due to differential global positioning system reading inaccuracy. O'Shaughnessy and Evett (2015) tested a center pivot system equipped with a commercial VRI package and noticed that (i) the width of edge effects between adjacent irrigation zones ranged from 1.52 m to 6.1 m and (ii) the mean length of the edge effect was 9.1 m in the travel direction of the pivot. The spatial resolution of VRI systems and therefore the minimum size of IMZ in practice are influenced by the edge effect, a fact that should be considered for VRI implementation. In a recent case study, Higgins et al. (2016) developed a methodology to determine IMZ size and reported 23 m × 23 m as the smallest size for an independent IMZ for a nine-span center pivot equipped with a commercial VRI package.

Mechanical characteristics of center pivot and lateral move irrigation systems impact water application patterns across a field. These systems are often called continuous move systems; yet, in reality, a series of stop-advance sequences occurs independently for each tower along the lateral when these systems operate. O'Shaughnessy et al. (2013) suggested future field trials to assess the effect of intermittent movement of inner towers on performance of their 6-span center pivot irrigation system. Chávez et al. (2010c) studied and modelled the inherent lateral move water application errors and developed a methodology to use VRI systems to compensate for those errors.

5.3 Challenges and opportunities for adoption of site-specific irrigation systems

Current uses of commercial site-specific VRI technologies are on a fairly coarse scale (Kranz et al., 2012), and adoption of these machines by growers has been slow to develop and hard to rationalize considering energy and water saving at current prices along with current high implementation and management costs (Evans et al., 2013). This is also accompanied by the fear of complexity and lack of experience associated with successful implementation of VRI technologies among growers and service providers along with limited reported research-based results on the impacts of VRI technologies on crop production.

The focus of this chapter and the main potential benefits of VRI technologies are to enhance irrigation WUE and improve crop production by maintaining soil water content within the optimum range across the field while minimizing run-off and deep percolation, and making the best use of stored water within the root zone (Hedley and Yule, 2009b). Other potential advantages of site-specific VRI technologies are (i) ease of site-specific chemigation/fertigation, (ii) avoiding irrigation of non-cropped areas (e.g., roads, ponds, ditches, farm tracks, drainage pathways and water ways), (iii) enhanced mixed cropping capability and (iv) applying liquid animal waste to specific areas in a field especially when there are regulatory limitations on amount and location of application to protect wells and to reduce leaching and contamination of wildlife areas (Kranz et al., 2012; Hedley et al., 2009).

A study on adoption of variable rate technology by cotton growers in Texas, USA, showed higher adoption rates for areas with higher field-level variability and reported higher likelihood of variable rate technology adoption with younger growers, growers with big operations and those who use computers in their farming operations (Nair et al., 2012). A recent study in Brazil showed that expansion of precision farming is related to field-observed agronomic and economic benefits (Borghi et al., 2016). A study on PA technology adoption by cotton growers in the United States revealed that belief in potential improvement in environmental quality and adoption of other PA technologies were the main factors correlated with the early adoption of three PA technologies: grid soil sampling, yield monitoring and RS (Watcharaanantapong et al., 2014).

Growers and policymakers are becoming more interested in SS-VRI mainly because of the increasing limitation on irrigation water supplies and environmental issues around the world (Evans et al., 2013). In Europe, the greatest short-term enhancement in WUE is projected to come from precision irrigation adoption of overhead irrigation systems (Monaghan et al., 2013). This is because a majority of irrigated lands in Europe utilize overhead irrigation systems (including center pivots, sprinklers and high-pressure rain guns) for which the commercial VRI technology has been developed. A variety of factors such as commodity price, water price and availability, water quality, degree of field-level soil heterogeneity and climatic conditions influence expected improvement in production quality and quantity by site-specific irrigation systems. This economic improvement will not be high enough in all situations to cover the cost associated with implementation and management of these systems. However, widespread adoption of site-specific irrigation systems may be beneficial at a global scale by enhancing the water footprint and environmental impacts of irrigated agriculture through improvement in field-level WUE and reduction in nitrogen leaching out of the root zone (King et al., 2006).

6 Future trends and conclusion

6.1 Development of decision support tools

PA is progressing with emerging innovations in instrumentation and measurements, a revolution that is changing the concept of an agricultural unit from farm to subfield and down to individual plants. In the future, most farmers in developed countries will have access to precision farming equipment and be collecting numerous site-specific geospatial data. Previous VRI-related research projects have mainly focused on (i) designing irrigation systems to spatially vary irrigation rates and (ii) developing software/hardware for guiding the VRI system (Pan et al., 2013). Future research efforts should focus on the development of novel approaches for affordable data acquisition as well as robust data processing. At the same time, reliable automated decision support tools should be developed for real-time processing of data and control of irrigation systems. This is important because most research efforts have focused on providing more information to growers who have limited time to process the information (Howell et al., 2012). Recent advances in the development of smartphone applications for irrigation management (e.g. Bartlett et al., 2015; Vellidis et al., 2016) open up new horizons for availability of information to growers for real-time site-specific irrigation management.

6.2 Site-specific deficit irrigation

Site-specific irrigation implementation involves extra irrigation system hardware, labour and site-specific data to detect spatio-temporal changes in soil water status and/or crop stress in each IMZ (King et al., 2006). More studies are needed on site-specific strategies for deficit irrigation using VRI technologies specifically for regions where water allocation restrictions are prescribed for irrigation and/or water charges exist. These constraints can provide enough economic motivation for growers to make investments towards advanced irrigation technologies.

6.3 High-resolution monitoring of soil water status and ET

Development of underground soil moisture sensors (Dong et al., 2013) is a good example of emerging innovations that can substantially advance growers' ability for high-resolution monitoring of field-level soil water status. In addition, given the recent progresses in RS-based ET mapping, for example, development of Google Earth Engine Evapotranspiration Flux, Allen et al. (2015), new tools are likely to be developed in the near future to generate affordable spatial ET maps with high enough spatio-temporal resolution for SS-VRI management. Such maps can be potentially used to delineate IMZ, establish SS-WPFs and create irrigation prescription maps throughout the season.

6.4 On-farm irrigation research

Traditional irrigation/agronomic research is typically conducted in controlled environments following statistical principles on small-plot trials over multiple sites and across several years on a property allocated to research. These research trials are time-consuming and labour-intensive and hence may become unrealistic for the near future (Drummond et al., 2003). In addition, scaling and transforming findings of small-plot research trials to operational management of growers' fields are challenging. On-farm irrigation experimentation through cooperation between researchers and growers becomes more viable with more trustable results using site-specific irrigation systems which help to develop site-specific irrigation solutions.

7 List of abbreviations

ANN	Artificial neural network
BD	Bulk density
ECa	Apparent electrical conductivity
ET	Evapotranspiration
FC	Field capacity
GIS	Geographical information system
GPS	Global positioning system
GNSS	Global navigation satellite system
GWR	Geographically weighted regression
ILP	Integer linear programming
IMZ	Irrigation management zone
K-NN	K-nearest neighbour

KG	Kriging
MLR	Multiple linear regression
MZ	Management zone
PA	Precision agriculture
PAW	Plant available water (also known as root zone available water holding capacity)
PTF	Pedotransfer function
PWP	Permanent wilting point
RKG	Regression kriging
RS	Remote sensing
SS-VRI	Site-specific variable rate irrigation
SS-WPF	Site-specific water production function
VRI	Variable rate irrigation
WPF	Water production function
WUE	Water use efficiency

8 Where to look for further information

- *Journals: Precision Agriculture* (publisher: Springer), *Computers and Electronics in Agriculture* (publisher: Elsevier), *Transactions of the ASABE* (publisher: American Society of Agricultural and Biological Engineers), *Irrigation Science* (publisher: Springer), *Irrigation and Drainage Engineering* (publisher: American Society of Civil Engineers), *Agricultural Water Management* (publisher: Elsevier).
- Books: Geostatistical Applications for Precision Agriculture (publisher: Springer).
- *Events organized by:* American Society of Agricultural and Biological Engineers (website: www.asabe.org), International Society of Precision Agriculture (website: www.ispag.org), tri-societies [American Society of Agronomy (website: www.agronomy.org), Crop Science Society of America (website: www.crops.org), and Soil Science Society of America (website: www.soils.org)].

9 References

Abdu, H., Robinson, D. A., Seyfried, M. and Jones, S. B. (2008). Geophysical imaging of watershed subsurface patterns and prediction of soil texture and water holding capacity. *Water Resources Research*, 44(4). W00D18, doi:10.1029/2008WR007043.

Allen, R., Morton, C., Kamble, B., Kilic, A., Huntington, J., Thau, D., et al. (2015). EEFlux: A Landsat-based Evapotranspiration mapping tool on the Google Earth Engine. In: *Proceedings of the Joint ASABE/IA Irrigation Symposium*. ASABE, St Joseph, MI, USA. doi:10.13031/irrig.20152143511

Archie, G. E. (1942). The electrical resistivity log as an aid in determining some reservoir characteristics. Transactions of the American Institute of Mining and Metallurgical Engineers, 146, 54–61.

Auerswald, K., Simon, S. and Stanjek, H. (2001). Influence of soil properties on electrical conductivity under humid water regimes. Soil Science, 166(6), 382–90.

Bartlett, A. C., Andales, A. A., Arabi, M. and Bauder, T. A. (2015). A smartphone app to extend use of a cloud-based irrigation scheduling tool. *Computers and Electronics in Agriculture*, 111, 127–30.

Boluwade, A., Madramootoo, C. and Yari, A. (2015). Application of unsupervised clustering techniques for management zone delineation: Case study of variable rate irrigation in Southern Alberta, Canada. Journal of Irrigation and Drainage Engineering, 142(1), 05015007.
Borghi, E., Avanzi, J. C., Bortolon, L., Junior, A. L. and Bortolon, E. S. (2016). Adoption and use of precision agriculture in Brazil: Perception of growers and service dealership. Journal of Agricultural Science, 8(11), 89.
Bouma, J. (1989). Using soil survey data for quantitative land evaluation. In *Advances in Soil Science* (pp. 177–213). New York, USA: Springer.
Brock, A., Brouder, S. M., Blumhoff, G. and Hofmann, B. S. (2005). Defining yield-based management zones for corn–soybean rotations. Agronomy Journal, 97(4), 1115–28.
Bronson, K. F., Booker, J. D., Bordovsky, J. P., Keeling, J. W., Wheeler, T. A., Boman, R. K., et al. (2006). Site-specific irrigation and nitrogen management for cotton production in the Southern High Plains. Agronomy Journal, 98(1), 212–19.
Camp, C. R., Sadler, E. J., Evans, D. E., Usrey, L. J. and Omary, M. (1998). Modified center pivot system for precision management of water and nutrients. Applied Engineering in Agriculture, 14(1), 23–31.
Chávez, J. L., Pierce, F. J., Elliott, T. V. and Evans, R. G. (2010a). A remote irrigation monitoring and control system for continuous move systems. Part A: Description and development. Precision Agriculture, 11(1), 1–10.
Chávez, J. L., Pierce, F. J., Elliott, T. V., Evans, R. G., Kim, Y. and Iversen, W. M. (2010b). A remote irrigation monitoring and control system (RIMCS) for continuous move systems. Part B: Field testing and results. Precision Agriculture, 11(1), 11–26.
Chávez, J. L., Pierce, F. J. and Evans, R. G. (2010c). Compensating inherent linear move water application errors using a variable rate irrigation system. Irrigation Science, 28(3), 203–10.
Cid-Garcia, N. M., Albornoz, V., Rios-Solis, Y. A. and Ortega, R. (2013). Rectangular shape management zone delineation using integer linear programming. Computers and Electronics in Agriculture, 93, 1–9.
Cohen, Jacob (1960). A coefficient of agreement for nominal scales. Educational and Psychological Measurement 20 (1): 37–46.
Córdoba, M., Bruno, C., Costa, J. and Balzarini, M. (2013). Subfield management class delineation using cluster analysis from spatial principal components of soil variables. Computers and Electronics in Agriculture, 97, 6–14.
Corwin, D. L. and Lesch, S. M. (2010). Delineating site-specific management units with proximal sensors. In *Geostatistical Applications for Precision Agriculture* (pp. 139–65). Dordrecht, Netherlands: Springer.
Corwin, D. L. and Lesch, S. M. (2003). Application of soil electrical conductivity to precision agriculture: Theory, principles, and guidelines. *Agronomy Journal*, 95, 455–71.
Debaene, G., Niedźwiecki, J., Pecio, A. and Żurek, A., (2014). Effect of the number of calibration samples on the prediction of several soil properties at the farm-scale. Geoderma, 214, 114–25.
Dong, X., Vuran, M. C. and Irmak, S. (2013). Autonomous precision agriculture through integration of wireless underground sensor networks with center pivot irrigation systems. Ad Hoc Networks, 11(7), 1975–87.
Drummond, S. T., Sudduth, K. A., Joshi, A., Birrell, S. J. and Kitchen, N. R. (2003). Statistical and neural methods for site-specific yield prediction. *Transactions of the ASAE*, 46(1), 5.
Dukes, M. D. and Perry, C. (2006). Uniformity testing of variable-rate center pivot irrigation control systems. *Precision Agriculture*, 7(3), 205.
Duncan, H. A., (2012). Locating the variability of soil water holding capacity and understanding its effects on deficit irrigation and cotton lint yield. Master's Thesis, University of Tennessee. http://trace.tennessee.edu/utk_gradthes/1286.
Earl, R., Taylor, J. C., Wood, G. A., Bradley, I., James, I. T., Waine, T., et al. (2003). Soil factors and their influence on within-field crop variability, part I: Field observation of soil variation. *Biosystems Engineering*, 84(4), 425–40.
Evans, R. G., LaRue, J., Stone, K. C. and King, B. A. (2013). Adoption of site-specific variable rate sprinkler irrigation systems. Irrigation Science, 31(4), 871–87.

Fortes, R., Millán, S., Prieto, M. H. and Campillo, C. (2015). A methodology based on apparent electrical conductivity and guided soil samples to improve irrigation zoning. Precision Agriculture, 16(4), 441–54.

Fraisse, C. W., Duke, H. R. and Heerman, D. F. (1995a). Laboratory evaluation of variable water application with pulse irrigation. Transactions of the ASAE, 38(5), 1363–9.

Fraisse, C. W., Heermann, D. F. and Duke, H. R. (1995b). Simulation of variable water application with linear-move irrigation systems. Transactions of the ASAE, 38(5), 1371–6.

Fraisse, C. W., Sudduth, K. A. and Kitchen, N. R. (2001). Delineation of site-specific management zones by unsupervised classification of topographic attributes and soil electrical conductivity. Transactions of the ASAE, 44(1), 155–66.

Haghverdi, A., Cornelis, W. M. and Ghahraman, B. (2012). A pseudo-continuous neural network approach for developing water retention pedotransfer functions with limited data. Journal of Hydrology, 442, 46–54.

Haghverdi, A., Leib, B. G., Washington-Allen, R. A., Ayers, P. D. and Buschermohle, M. J. (2015a). High-resolution prediction of soil available water content within the crop root zone. Journal of Hydrology, 530, 167–79.

Haghverdi, A., Leib, B. G., Washington-Allen, R. A., Ayers, P. D. and Buschermohle, M. J. (2015b). Perspectives on delineating management zones for variable rate irrigation. *Computers and Electronics in Agriculture*, 117, 154–67.

Haghverdi, A., Leib, B. G., Washington-Allen, R. A., Buschermohle, M. J. and Ayers, P. D. (2016). Studying uniform and variable rate center pivot irrigation strategies with the aid of site-specific water production functions. *Computers and Electronics in Agriculture*, 123, 327–40.

Haghverdi, A., Öztürk, H. S. and Cornelis, W. M. (2014). Revisiting the pseudo continuous pedotransfer function concept: Impact of data quality and data mining method. *Geoderma*, 226, 31–8.

Han, Y. J., Khalilian, A., Owino, T. O., Farahani, H. J. and Moore, S. (2009). Development of Clemson variable-rate lateral irrigation system. *Computers and Electronics in Agriculture*, 68(1), 108–13.

Hedley, C. B. and Yule, I. J. (2009a). A method for spatial prediction of daily soil water status for precise irrigation scheduling. *Agricultural Water Management*, 96(12), 1737–45.

Hedley, C. B. and Yule, I. J. (2009b). Soil water status mapping and two variable-rate irrigation scenarios. *Precision Agriculture*, 10(4), 342–55.

Hedley, C. B., Yule, I. J., Tuohy, M. P. and Vogeler, I. (2009). Key performance indicators for simulated variable-rate irrigation of variable soils in humid regions. *Transactions of the ASABE*, 52(5), 1575–84.

Herbst, M., Diekkrüger, B. and Vereecken, H., (2006). Geostatistical co-regionalization of soil hydraulic properties in a micro-scale catchment using terrain attributes. *Geoderma*. 132(1), 206–21.

Higgins, C. W., Kelley, J., Barr, C. and Hillyer, C. (2016). Determining the minimum management scale of a commercial variable-rate irrigation system. *Transactions of the ASABE*, 59 (6): 1671–80.

Howell, T. A., Evett, S. R., O'Shaughnessy, S. A., Colaizzi, P. D. and Gowda, P. H. (2012). Advanced irrigation engineering: Precision and Precise. *Journal of Agricultural Science and Technology. A*, 2(1A), 1–9.

Iqbal, J., Thomasson, J. A., Jenkins, J. N., Owens, P. R. and Whisler, F. D. (2005). Spatial variability analysis of soil physical properties of alluvial soils. *Soil Science Society of America Journal*, 69(4), 1338–50.

Karmeli, D. and Peri, G. (1974). Basic principles of pulse irrigation. *Journal of the Irrigation and Drainage Division*, 100(3), 309–19.

Kerry, R., Oliver, M. A. and Frogbrook, Z. L. (2010). Sampling in precision agriculture. In *Geostatistical Applications for Precision Agriculture*, ed. M. A. Oliver (pp. 35–63). Dordrecht, Netherlands: Springer.

Kim, Y., Evans, R. G. and Iversen, W. M. (2009). Evaluation of closed-loop site-specific irrigation with wireless sensor network. *Journal of Irrigation and Drainage Engineering*, 135(1), 25–31.

King, B. A., Brady R. A., McCann I. R. and Stark J. C. (1995). Variable rate water application through sprinkler irrigation. In *Site-specific Management for Agricultural Systems*, eds. P. C. Robert, R. H. Rust and W. E. Larson (pp. 485–93). Madison, WI, American Society of Agronomy.

King, B. A., Stark, J. C. and Wall, R. W. (2006). Comparison of site-specific and conventional uniform irrigation management for potatoes. *Applied Engineering in Agriculture*, 22(5), 677–88.

King, B. A., Wall, R. W., Kincaid, D. C. and Westermann, D. T. (2005). Field testing of a variable rate sprinkler and control system for site-specific water and nutrient application. *Applied Engineering in Agriculture*, 21(5), 847–54.

King, B. and Kincaid, D. C. (2004). A variable flow rate sprinkler for site-specific irrigation management. *Applied Engineering in Agriculture*, 20(6), 765–70.

Kitchen, N. R., Sudduth, K. A., Myers, D. B., Drummond, S. T. and Hong, S. Y. (2005). Delineating productivity zones on claypan soil fields using apparent soil electrical conductivity. *Computers and Electronics in Agriculture*, 46(1), 285–308.

Kranz, W. L., Evans, R. G., Lamm, F. R., O'Shaughnessy, S. A. and Peters, R. T. (2012). A review of mechanical move sprinkler irrigation control and automation technologies. *Applied Engineering in Agriculture*, 28(3), 389–97.

Liang, X., Liakos, V., Wendroth, O. and Vellidis, G. (2016). Scheduling irrigation using an approach based on the van Genuchten model. *Agricultural Water Management*, 176, 170–9.

Monaghan, J. M., Daccache, A., Vickers, L. H., Hess, T. M., Weatherhead, E. K. and Grove, I. G., et al. (2013). More 'crop per drop': Constraints and opportunities for precision irrigation in European agriculture. *Journal of the Science of Food and Agriculture*, 93(5), 977–80.

Moral, F. J., Terrón, J. M. and Da Silva, J. M. (2010). Delineation of management zones using mobile measurements of soil apparent electrical conductivity and multivariate geostatistical techniques. *Soil and Tillage Research*, 106(2), 335–43.

Mouazen, A. M., Alhwaimel, S. A., Kuang, B. and Waine, T. (2014). Multiple on-line soil sensors and data fusion approach for delineation of water holding capacity zones for site specific irrigation. *Soil and Tillage Research*, 143, 95–105.

Nair, S., Wang, C., Segarra, E., Johnson, J. and Rejesus, R. (2012). Variable rate technology and cotton yield response in Texas. *Journal of Agricultural Science and Technology. B*, 2(9B), 1034.

O'Shaughnessy, S. A. and Evett, S. R. (2015). Zone edge effects with variable rate irrigation. Proceedings of the 27th Annual Central Plains Irrigation Conference, Colby, KS, USA. Available online at: https://www.ksre.k-state.edu/irrigate/oow/p15/OShaughnessy_15.pdf.

O'Shaughnessy, S. A., Evett, S. R. and Colaizzi, P. D. (2015). Dynamic prescription maps for site-specific variable rate irrigation of cotton. *Agricultural Water Management*, 159, 123–38.

O'Shaughnessy, S. A., Urrego, Y. F., Evett, S. R., Colaizzi, P. D. and Howell, T. A. (2013). Assessing application uniformity of a variable rate irrigation system in a windy location. *Applied Engineering in Agriculture*, 29(4), 497–510.

Omary, M., Camp, C. R. and Sadler, E. J. (1997). Center pivot irrigation system modification to provide variable water application depths. *Applied Engineering in Agriculture*, 13(2), 235–39.

Pan, L., Adamchuk, V. I., Martin, D. L., Schroeder, M. A. and Ferguson, R. B. (2013). Analysis of soil water availability by integrating spatial and temporal sensor-based data. *Precision Agriculture*, 14(4), 414–33.

Rivers, E.D. and Shipp R.F. (1972). Available water capacity of sandy and gravelly North Dakota soil. *Soil Science*, 113, 74–80.

Romano, N. and Santini, A. (2002). Field, In: *Methods of Soil Analysis*. Part 4, Physical Methods, Soil Sci. Soc. Am. Book Ser., vol. 5, edited by J. H. Dane and G. C. Topp (pp. 721–38). Madison, WI, USA: Soil Science Society of America.

Roth, R. L. and Gardner, B. R. (1989). Modified self-moving irrigation system for water-nitrogen crop production studies. *Applied Engineering in Agriculture* 5(2): 175–9.

Sadler, E. J., Evans, R., Stone, K. C. and Camp, C. R. (2005). Opportunities for conservation with precision irrigation. *Journal of Soil and Water Conservation*, 60(6), 371–8.

Saey, T., Van Meirvenne, M., Vermeersch, H., Ameloot, N. and Cockx, L. (2009). A pedotransfer function to evaluate the soil profile textural heterogeneity using proximally sensed apparent electrical conductivity. *Geoderma*, 150(3), 389–95.

Sudduth, K. A., Kitchen, N. R., Wiebold, W. J., Batchelor, W. D., Bollero, G. A., Bullock, D. G. and Thelen, K. D. (2005). Relating apparent electrical conductivity to soil properties across the north-central USA. *Computers and Electronics in Agriculture*, 46(1), 263–83.

Sui, R. and Fisher, D. K. (2015). Field test of a center pivot irrigation system. *Applied Engineering in Agriculture*, 31(1), 83–8.

Sui, R., Fisher, D. K. and Reddy, K. N. (2015). Yield response to variable rate irrigation in corn. *Journal of Agricultural Science*, 7(11), 11.

Van Genuchten, M.Th., (1980). A closed-form equation for predicting the hydraulic conductivity of unsaturated. *Soil Science Society of America Journal*, 43, 892–98.

Vellidis, G., Liakos, V., Andreis, J. H., Perry, C. D., Porter, W. M., Barnes, E. M., et al. (2016). Development and assessment of a smartphone application for irrigation scheduling in cotton. *Computers and Electronics in Agriculture*, 127, 249–59.

Vories, E., Rhine, M. and Straatmann, Z. (2016). Investigating irrigation scheduling for rice using variable rate irrigation. *Agricultural Water Management*. 179, 314–23.

Watcharaanantapong, P., Roberts, R. K., Lambert, D. M., Larson, J. A., Velandia, M., English, B. C., et al. (2014). Timing of precision agriculture technology adoption in US cotton production. *Precision Agriculture*, 15(4), 427–46.

Yari, A., Madramootoo, C. A., Woods, S. A. and Adamchuk, V. I. (2017). Performance evaluation of constant versus variable rate irrigation. *Irrigation and Drainage*. doi:10.1002/ird.2131

Zhang, X., Shi, L., Jia, X., Seielstad, G. and Helgason, C. (2010). Zone mapping application for precision-farming: A decision support tool for variable rate application. *Precision Agriculture*, 11(2), 103–14.

Chapter 2

Deficit irrigation and site-specific irrigation scheduling techniques to minimize water use

Susan A. O'Shaughnessy, USDA-ARS, USA; and Manuel A. Andrade, Oak Ridge Institute for Science and Education, USA

1   Introduction
2   DI strategies: overview
3   DI strategies: approaches, risks and advantages
4   SSIM: achieving precision irrigation
5   Variable rate irrigation
6   Integration of plant feedback sensor systems for site-specific VRI control
7   Conclusions
8   Where to look for further information
9   Acknowledgements
10 Disclaimer
11 References

1 Introduction

With water scarcity being prevalent on a wide scale, efficient use of water for agriculture is paramount to sustaining food security and meeting the demands for food, fibre, feed and fuel production. Advanced irrigation technologies and their adoptions, namely the conversion from gravity-flow systems to pressurized irrigation systems (sprinkler and micro-irrigation), have led to decreased water use (Pláyan and Mateos, 2006; NASS, 2013; Tarjuelo et al., 2015). Yet, regulations to increase water conservation from political bodies and competing economic sectors and, in some cases, declining water supplies still loom. Two important technologies to help minimize agricultural water use include deficit irrigation (DI) management and precision irrigation. Both methods entail particular strategies to manage the amount and timing of water applied for sustainable crop production for the present and future. Advancements in technology and irrigation scheduling strategies are key to improved agricultural water use, and simultaneously can help to improve environmental sustainability (Hedley, 2014; Fan and Brzeska, 2016). This chapter describes the main DI strategies used in agriculture and reports on results from current studies using DI strategies, as well as the status for site-specific irrigation management (SSIM) and its role in minimizing agricultural water use.

2 DI strategies: overview

When water supplies are limited, a farmer's goal should be to maximize net income per unit of water used rather than per land unit (Fereres and Soriano, 2006). One approach to maximizing economic yield per unit of water used by the crop (evapotranspiration, ET) is to practice some form of DI. A broad definition of DI is the practice of applying water in an amount that is less than needed to meet full ET requirements of a healthy plant without significantly affecting yield quantity or quality.

In practice, successfully managing profitable crop production with DI strategies is much more involved than simply reducing the amount of water applied. DI strategies require precise knowledge of crop response to water as drought tolerance varies considerably by species, cultivar and stage of growth (FAO, 2002). Effective DI requires a planned soil water depletion scheme (Howell et al., 2012) and, therefore, not only are crop physiological responses important, but soil water-holding characteristics as well. Early studies critical to successful DI management were those that documented plant response to limited plant available water, which began in journal publications in the 1950s. These include paired observations of crop status with varying levels of irrigation treatments, and measurements of decreased yield components (Robins and Domingo, 1953, 1956a,b, 1962). Hsiao (1973, 1974) and Chaves et al. (2002) described plant responses to water stress that aid in limiting transpiration. These include stomatal conductance, changes in leaf temperature, cell growth and metabolic variations. Detection of these changes, either through invasive means or through remote sensing, have been used to signal irrigations. Importantly, several published studies report that mild water stress (depending on the crop) could result in water savings without significantly affecting yield quantity or quality.

If DI is treated as an optimization strategy as suggested by Geerts and Raes (2009), then the goal to maximize crop water productivity (CWP) or yield per unit of water applied can be represented as:

$$CWP = Y/ET \qquad (1)$$

where Y is economic yield and ET is seasonal evapotranspiration, which represents the inputs of applied irrigation, 'effective' precipitation and soil water depletion from the root zone (Howell, 2001). Methods for estimating maximum crop yields using empirical production functions are found in *Irrigation and Drainage Paper No. 33 Yield Response to Water* (Doorenbos and Kassam, 1979), and have now been updated in *Irrigation and Drainage Paper No. 66* (Steduto et al., 2012), which also introduces the crop simulation model AquaCrop. Maximum values of CWP can be characterized by developing crop water production functions. Production functions can vary by climate, examples include functions reported by Grassini et al. (2009) for sunflower, by Klocke et al. (2011) for corn and by soil type as shown by Tolk and Howell (2012) for sunflower. Water productivity (CWP) may increase under DI, but it should be noted that maximum CWP generally occurs close to maximum yield as shown for late and early maturing sorghum hybrids grown in 2009 and 2010 (Fig. 1) (O'Shaughnessy et al., 2014). Furthermore, under extreme

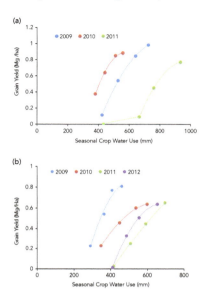

Figure 1 (a) Crop water production functions for a late maturing sorghum hybrid P84G62 and (b) early maturing sorghum hybrid NC+5C35 grown in 2009, 2010 and 2011 at Bushland, Texas.

to exceptional drought conditions, DI management is not a good strategy to improve CWP. As demonstrated for sorghum grown under the exceptional drought conditions of 2011 in Bushland, Texas, irrigation amounts that were less than 100% ET resulted in significantly reduced grain yields (see data for 2011, Fig. 1a) even though the irrigation amount applied was double that applied in 2009. Greater irrigation amounts for the early maturing sorghum hybrid, NC+5C35, were also required to meet atmospheric demands of moderate drought conditions in 2012 (Fig. 1b). Climate variability may be the single most important determinant of agricultural productivity through its effects on temperature and water regimes (Kang et al., 2009).

Although DI strategies are not prudent for growing crops under extreme and exceptional drought conditions (high winds, high air temperature and minimal rainfall), applying less than full irrigation amounts are promising methods to save water under more moderate atmospheric conditions. DI strategies can be achieved using various approaches: (1) limiting water throughout the irrigation season (sustained DI); (2) applying deficit water amounts at specific growth stages of a crop, regulated DI (RDI); (3) partial root zone drying (PRD); and (4) supplemental irrigation (SI). Improved water use efficiency is due mainly to enhanced guard cell signal transduction network that decreases water loss through transpiration, optimized stomatal control that improves the photosynthesis to transpiration ratio and decreased evaporation from wetted surface areas using PRD (Chai et al., 2016).

CWP is influenced by yield response to irrigation. Critical growth periods for water deficit can be different for different crops. Therefore, farmers and crop consultants must familiarize themselves with the interactions of irrigation management and crop response for each crop species. Many yields are affected during the anthesis (flowering) and pod, grain or boll filling stage as summarized by Howell (1990). According to Fereres and Soriano (2006), DI has had significantly more success in tree crops and vines than in field crops because economic return in tree crops are often associated with crop quality and are not directly related to biomass production and water use. In addition, yield-determining processes in many fruit trees are not sensitive to water deprivation at some developmental stages. However, DI strategies have been successfully implemented for grain, oil and vegetable production.

3 DI strategies: approaches, risks and advantages

3.1 Sustained deficit irrigation

Sustained deficit irrigation (sustained DI) research trials, where the amount of water applied is limited throughout the season, generally compare yield and CWP responses at different irrigation treatment levels. Farmers can use published results as a starting point for optimal crop water management practices. For example, Kifle and Gebretsadikan (2016) used different DI levels on potatoes: those irrigated at constant deficit of 25% less than full ET requirement suffered from significantly lower yields; however, applying 65% of full irrigation throughout the season produced better yields than both the 25% constant deficit and a regulated deficit applied at flowering. Camargo et al. (2015) found that potatoes irrigated at 80% of full ET requirements resulted in CWP levels that were similar or significantly better than potatoes irrigated at full ET. For camelina, an oil crop, DI at levels that were 60% and 70% of the fully irrigated crop produced yields that were 78% and 87% of maximum (Hergert et al., 2016). Sustained DI was also successfully implemented for sunflower: Garofalo and Rinaldi (2015) reported that yields of sunflower with small (150 mm) and moderate (270 mm) amounts of water ensured satisfactory seed production with water savings of 74% and 53%, respectively, compared with a well-watered treatment in a water-limited environment. Howell et al. (2004), O'Shaughnessy and Evett (2010a), Sampathkumar et al. (2013) and Lokhande and Reddy (2014) reported that cotton irrigated at less than the full ET rates (20–25% less) produced yields higher than plants irrigated to meet full ET requirements. Although the physiological method is not clear, some drought-tolerant corn hybrids produced higher CWP levels under DI as reported by Aydinsakir et al. (2013), Hao et al. (2015a) and Mounce et al. (2016). The irrigation amounts were approximately 75% of full ET.

3.2 Regulated deficit irrigation

A second DI scheme is RDI, which was introduced by Chalmers et al. (1986) to control vegetative growth in peach trees. Serendipitously, they discovered that the strategy did not impact fruit yield. With RDI, limited amounts of water are applied to a crop during stages when the crop is not sensitive to water stress, while full irrigation is applied during growth stages when yields can be significantly compromised by water deficits. Horticultural benefits of RDI (greater water use efficiency, less pruning, less incidence of disease, improved fruit quality, phenolic composition and higher sugar concentrations) have been documented for several other high value crops such as grapes, fruit trees and nut trees (Goldhamer and Salinas, 2000; Costello and Patter, 2012; Romero et al., 2013; Ballester et al., 2014; Lanari et al., 2014; Nangare et al., 2016) irrigation management is difficult without use of a reliable plant-based measure of stress (Shackel, 2011).

There are a number of plant-based sensors available for sensing water stress. These sensors can be divided into two main categories: those that measure tissue water status (e.g. pressure chambers for stem leaf water potential assessment; psychrometers; linear placement transducers) and those that measure physiological responses (e.g. dendrometers, sap flow gauges, infrared thermometers) (Jones, 2004). De la Rosa et al. (2015) measured fluctuations in trunk diameter for RDI scheduling on nectarine trees during a three-year study. The scheme involved irrigating at 60% less than full ET at post-harvest after the first fruiting season. They used linear variable displacement transducers to monitor trunk diameter fluctuation and midday stem water potential measurements to signal irrigations and maintain the signal intensity of the maximum daily shrinkage of the trunks of nectarine trees. This method resulted in a savings of 17, 15 and 37% of water over three seasons as compared with irrigation scheduling using 110% of crop ET. Girón et al. (2015) reported that trunk growth rates rather than maximum daily shrinkage measurements provided a useful indication of when to irrigate using a RDI management on olive trees. As a final example, Zhang and Wang (2013) used infrared thermometers to measure canopy temperature and calculated a thermal stress index to apply deficit amounts of water post-harvest to peach trees using furrow and subsurface drip irrigation (SDI). The greatest CWP resulted from deficit SDI as compared with full irrigation using furrow and with DI using furrow.

3.3 Partial root zone drying

PRD is a DI method that enables watering of only a portion of the root system, leaving the other portion to dry to a predetermined level before the next irrigation. Kuskabe et al. (2016) reported that PRD management saved 43–47% of water applied to grapefruit trees over two irrigation seasons without compromising fruit and juice quality. Spreer et al. (2007) demonstrated that fruit size in mango increased and fruits had a higher fraction of edible parts under PRD as compared with RDI management. Another benefit of PRD compared with RDI is improved nitrogen uptake by plants due to the increase in nutrient availability in the root zone (Wang et al., 2013; Liu et al., 2015). With today's technological advancements in soil water and plant sensors, crop simulation models and weather forecasting, farmers currently practising DI strategies should have a higher confidence level of success in maximizing CWP. In a review of DI practices, Chai et al. (2016) listed the water savings reported under PRD management from various countries for various crops to range from 11 to 54% of full irrigation.

3.4 Supplemental irrigation

Recently, frequent episodes of drought have pushed the fourth type of DI management method, deficit SI, to the forefront. SI is defined as 'the addition of small amounts of water to essentially rainfed crops during times when the rainfall fails to provide sufficient moisture for normal plant growth, in order to improve and stabilize yields' (Oweis, 1997, cited in Oweis and Hachum, 2009). Rainfed agriculture is the predominant agricultural production system worldwide (FAO, 2011). Strategies for optimal SI have been investigated for some time in various regions for dry land crop production (Kurian, 1975; Madramootoo and Jutras, 1984; Carsky et al., 1995) and to a limited degree are being explored in some parts of the United States (Boyer et al., 2014). Because many of these regions are also experiencing increasing water scarcity and variable seasonal rainfall with periods of drought, practising deficit SI management is critical.

Similar to non-supplemental irrigation practices, a good starting point for determining deficit SI management practices (irrigation amounts and timing) is to use crop production functions. However, these production functions should be based on SI amounts and the probability of seasonal rainfall and crop yields for the region (Oweis and Hachum, 2009). A study by Wang et al. (2013) reported on the effect of deficit amounts and different irrigation timings of SI on wheat yields. The optimal deficit SI treatment was determined to be maintenance of soil water content at 75% of field capacity and applying two irrigations, that is, when the 4th node was detectable and after anthesis was complete. Guo et al. (2015) reported similar results for deficit SI management of wheat, but managed soil water content in the top 40 cm at 70% of field capacity. Assefa et al. (2016) harvested rainwater and applied deficit SI to onions during periods of long dry spells and overall low rainfall. Yields irrigated at 75% of ET using SI management were not significantly different compared with yields irrigated at 100% of ET. In a Mediterranean environment, Karrou and Oweis (2012) showed that deficit SI at 33% of full for wheat and 67% of full for legume produced optimal CWP compared with full SI. Rey et al. (2016) reported that deficit SI can also be key to achieving profitable crop production in humid climates during drought years. Although water savings can be achieved by using deficit SI practices, the relationship between production prices (when high) and irrigation costs (low) may not incentivize farmers to use deficit SI management (Oweis and Hachum, 2009).

3.5 Use of drought-tolerant cultivars

Increased acreage to corn production and occurrences of drought in the United States have also been an impetus behind the development of drought-tolerant hybrids (Roth et al., 2013). Drought-tolerant crops can have multiple drought-resistant traits, which can interact with one another and with environmental factors and crop management practices (Sinclair, 2011; Fracasso et al., 2016). Matching the key phenotypic traits for drought resistance with the different DI strategies can help optimize CWP. For example, applying deficit amounts of water during the vegetative period, rather than the reproductive stages, for crops that resist drought by using less water through stomatal regulation and shorter plant height, could facilitate improved CWP (Hao et al., 2015b; Mounce et al., 2016). On the other hand, Gheysari et al. (2017) demonstrated that CWP in drought-tolerant maize grown for silage benefited more from continuous DI management over the growing season as compared with RDI management. The drought-tolerant traits were greater rooting density and depth. However, for maize grain yields, Vadez (2014) stressed that deep-water extraction, due to an extensive root system, may be heightened by irrigation applied at the reproductive stages. More studies need to be performed to indicate the optimal DI strategy for emerging drought-tolerant hybrids.

3.6 Risks of DI

Clearly, DI management comes with risks. For example, severe water stress and water stress coupled with heat stress in the reproductive stages of grain crops could result in significant deleterious effects such as reductions in root development (Sharpe and Davies, 1979), leaf area and canopy height (Denmead and Shaw, 1960), pollen sterility, kernel abortion (Hays et al., 2007) or total crop loss (Benjamin et al., 2015). Even moderate water stress conditions affected protein production in grain crops (Aydinsakir et al., 2013; Ertek and Kara, 2013; and Tari, 2016). DI should not be practised during severe or extreme drought years (O'Shaughnessy et al., 2014), which can significantly reduce yields even in drought-tolerant species. Another caveat is that reducing irrigation water use by practising DI reduces leaching and thereby increases the risk of soil salinity. An increase in soil salinity decreases the potential for maximum CWP (Fereres and Soriano, 2006). Aragüés et al. (2014) report that DI strategies increase soil salinization and sodification when irrigated with low-quality waters. Soil salinity, chloride concentration (Cle) and sodicity (SARe) generally increased after the irrigation season, more so in the sustained DI treatment. However, they reported that water use was reduced by 40% compared with full irrigation and water productivity increased by 65% without adverse effects in peach tree response (i.e. yields, productivity and Na and Cl toxicity symptoms). In question is the long-term sustainability of DI practices with saline water as salt-affected soils adversely affect plant growth and yields, and many aspects of microbial activity (Setia et al., 2011; Yan et al., 2015). Studies have consistently shown a build-up of salts near emitters and transient saline-sodic conditions (Mounzer et al., 2013). Irrigation with saline water (including leaching practices) inevitability increases salinity loads in soils to levels that could significantly reduce crop yield (Acosta et al., 2011) and render soils unusable (20Datta and Jong, 2002; Wichelns and Qadir, 2015). Adequate salt, irrigation and drainage management are imperative to sustain crop production under DI strategies and especially with low-quality water.

3.7 Virtual blue water savings

Optimizing irrigation management with DI strategies not only provides for gain in crop yield with less water, but for water savings to be used for irrigated production on more acreage, as water embedded in agricultural goods for interstate or international trade (Biewald et al., 2014), or for use by other economic sectors. This concept, often referred to as virtual blue water (surface and groundwater) savings, has regained attention in discussions aimed at providing ample water on a national scale (Liu et al., 2015) and meeting global demands for agricultural products while compensating for freshwater and agricultural water deficits. Biewald et al. (2014) reported that countries in the Middle East and South Asia will profit from trade by importing water-intensive crops. However, Perry (2014) argued that water consumption does not matter globally, and Boelens and Vos (2012) argued that international food trade can seriously affect poor consumers as they become dependent on cheap imported food, and poor producers cannot compete with the imported food. For regions that export water-intensive crops, there continues to be an economic advantage to maximize yields because the cost of irrigated water does not reflect its scarcity. Zhang and Anadon (2014) used this same reasoning in discussing unsustainable water use patterns within China. Although there is the potential to gain from water saved using DI strategies, the economic advantages and policies to effect real solutions of water savings are not always in place.

4 SSIM: achieving precision irrigation

Early irrigators practised SSIM with clay pots located below the soil surface (Sheng Han, 1974 as cited by Bainbridge, 2001) and above surface SSIM with buckets. Even though these were low-tech systems, farmers were able to benefit from water savings and increased

CWP (Batchelor et al., 1996). SSIM or precision irrigation requires irrigation application methods that enable a specific volume of water to be applied to an exact location within a field. Optimal management of drip or trickle irrigation systems may be labelled precision irrigation, while other researchers refer only to variable rate irrigation (VRI) as precision irrigation (Daccache et al., 2015). Meeting the specifications of precise amounts of water applied to precise locations can be achieved by zone-controlled drip irrigation systems (more likely to be used in horticulture, Zude-Sasse et al., 2016), precision mobile drip irrigation (PMDI) systems and moving VRI systems (centre pivot or linear move).

Recently, PMDI has been introduced into the Central and Southern High Plains regions as a system to improve crop water use efficiency. Early forms of PMDI are reported in the literature as travelling trickle systems (Rawlins et al., 1974; Phene et al., 1985). A PMDI system adapts drip irrigation hose onto a moving irrigation system (Fig. 2), with the goal to apply water uniformly in the direction of sprinkler travel and efficiently at the soil surface, eliminating run-off and water losses from sprinkler spray application. When row crops are planted in circles using a GPS-guided tractor, the drops can be aligned in the centre of the furrows. The additional horizontal and vertical guide-wire bracing helps to maintain the drip line in the centre of the furrow and limit its tendency to climb onto the crop, which could reduce the system's application efficiency. A number of different design methods to outfit the drip line onto the pipeline exist, including replacing the 1.91-cm drop hose with an ultraviolet PVC sleeve, sliding a PVC sleeve over the 1.91-cm drop and adding a low-drift nozzle sprinkler package using flex hose or plumbed-in using PVC hardware with a 90 elbow. Vertical guide wires attached to a 16 gauge horizontal cable are used to secure the drop lines in place (Fig. 2).

The PMDI system meets precision application criteria in a spatial sense because of its capability to apply a precise amount of water within a precise location, that is, between plant rows. Olson and Rogers (2008) compared corn yields grown in northwest Kansas with PMDI and low elevation spray application (LESA), where the nozzles were at heights approximately 18 inches above the ground. They reported no yield differences. Recently, the specialized drip line has been modified to incorporate pressure compensated emitters. With emitter spacing of six inches, the capacity of the drip line is one gallon per hour per foot. Kisekka et al. (2016) also reported no significant difference in yields compared with

Figure 2 Precision mobile drip irrigation adapted onto a centre pivot system. Photo taken of a precision mobile drip irrigation system in Bushland, Texas.

LESA application over corn in central Kansas, but they do report other benefits of PMDI, such as maintaining dry wheel tracks and sustaining irrigated agriculture on farms where well capacities are low. However, to meet full crop water needs, the moving sprinkler may need to irrigate continuously throughout the growing season.

5 Variable rate irrigation

VRI systems provide flexibility in managing water application rates to address variability in crop water stress across a field. The variability may be due to differences in topography, soil textures, available water-holding capacity and soil electrical conductivity. VRI is also used to manage non-uniform crop conditions, due to drainage problems, variability in plant density due to poor emergence, uneven distribution of residue and pest infestation or disease (Grisso et al., 2009). These characteristics can be determined remotely (Hatfield and Prueger, 2010) and used to modify watering levels if plant density or plant vigour is significantly low. Today's commercial VRI systems have been tested and shown to perform reliably (Han et al., 2009; O'Shaughnessy et al., 2013; Sui and Fisher, 2015). The design of sprinkler irrigation systems is based on fundamental engineering principles. Over the years, growers in different regions of the United States have adopted specific sprinkler application methods to afford the most cost-effective design and/or improve irrigation application efficiencies. For example, sprinkler spacing in semi-arid regions is typically five feet (1.5 m) and growers have adopted low energy precision application (LEPA) (bubblers or drag socks located near on the ground) (Lyle and Bordovsky, 1983) or LESA to help reduce water losses from evaporation and wind. Application methods using LESA in the Texas High Plains region is more common than LEPA methods. While in humid regions, it is common to find sprinkler spacing nearly twice that distance (at 9–18 feet apart) and use of mid-elevation spray application with nozzles located approximately five feet from the ground (Gossel et al., 2013).

With VRI centre pivot systems, there are three methods to accomplish VRI: (1) speed or sector control, which refers to the variable application of irrigation amounts in the direction of pivot travel only by changing the pivot speed; (2) zone control, which enables variable irrigation applications along the lateral pipeline and in the direction of travel of the moving sprinkler system; and (3) individual sprinkler zone control. LaRue and Frederick (2012) discussed steps that can be taken to determine which type of system best suits a farmer's needs and which system will result in a lesser return of investment. Although commercial VRI systems are reliable, questions remain how to use effectively such a system, including drip and PMDI systems, to address within-field spatial variation. A good starting point is to develop a map of the field and locate areas that should not receive water such as roads, rock outcroppings, ponds and streams. In the case of zone control, appropriate sprinkler zones can be turned off over these areas. In situations where excessive slopes exist or low areas in a field tend to pond water, the speed of the VRI system can be increased (for zone control, the watering rate can be reduced), to decrease the potential for run-off or waterlogging, respectively.

Research studies have shown that timely data streams from aerial imagery (DeTar et al., 2006; Gonzalez-Dugo et al., 2013; Agam et al., 2014), weather data, remote ground-based plant and *in situ* soil water sensors provide decision support for VRI management. However, inclusion of the element of timing (i.e. when to apply an irrigation) requires the incorporation of feedback from sensor network systems. It is not difficult to realize that these high-tech machines with sensor network system require decision support software to help manage the site-specific VRI systems in a timely and prudent manner. The timing component (when to irrigate) of SSIM systems is possible with the convergence of affordable computing equipment, reliable wireless communication protocols and robust sensor network systems. The sensor network systems that show potential are those that provide continuous weather, plant and soil water measurements and are used to develop decision support or recommendations for improved crop water management.

While the addition of sensor network systems complicates the operation of precision irrigation systems, the complexities can be overcome with a software system that automates its operation, including centralized data management, data processing and provision of easily understandable decision support recommendations for whole-field irrigation management. The software must be able to provide a graphical user interface that is easy to access and shows readily discernible information. McCarthy et al. (2010) introduced a simulation framework to aid in the development, evaluation and management of spatially and temporally varied site-specific irrigation control strategies. The core of the algorithm is a cotton model that uses inputs of soil properties (specific to each management zone (MZ) within a field), plant sensing data and weather data to simulate yield and water use efficiency based on a combination of the inputs. Thorpe et al. (2015) described the use of a geographic information system (GIS) to remotely sensed data and the CSM-CROPGRO-Cotton model for post-analysis of spatial cotton yield, water use and irrigation requirements. Andrade et al. (2015) provided a summary of a GIS-based system that manages a VRI centre pivot system in near real-time as well as performs post-harvest analysis. This system is discussed in more detail later in the chapter.

Current protocols for SSIM require that a field be divided into homogeneous areas or MZs. MZs have been typically characterized using apparent electrical conductivity (EC_a) data (Hedley and Yule, 2009; Vories et al., 2017), slope and elevation changes (Evans et al., 2013) and soil textural characteristics (Stone et al., 2016). These zones do not need to be contiguous but they should contain similar characteristics and be managed consistently. Irrigation management for each zone is controlled by a prescription map, which prescribes the watering level for each MZ. The precise volume of water to be applied to a MZ is an estimate of crop water needs that could be based on daily soil water status maps (Hedley and Yule, 2009), peak daily water demand (O'Shaughnessy et al., 2015) or software decision support tools supplemented with local data such as rainfall, irrigation amounts and soil water potential readings as with the use of IrrigatorPro (Stone et al., 2016).

To incorporate the element of time or the change in crop water needs over an irrigation season, SSIM must be able to detect differences in crop water status across a field and as the crop matures. To accomplish this, VRI systems must be outfitted with sensor network systems to assess temporal differences in plant water status or in plant available soil water content. For large-sized fields, sensing network systems and aerial imagery (Berni et al., 2009; Rosenberg et al., 2013) could help to overcome infeasible time constraints that are required for monitoring and maintaining a well-managed field. Large farms were the first to invest and adopt advanced VRI technology (Milton et al., 2006). The integration of sensing network systems with VRI systems provides continuous spatial and temporal information that can strengthen precision irrigation scheduling practices.

The addition of sensor network systems provides a means for timely crop water management. Sensor network systems comprised of wireless sensors and smart nodes that offer remarkable opportunities for the development and application of real-time management system for site-specific irrigation and hardware control (Coates and Delwiche, 2008; Kim et al., 2008, 2009; O'Shaughnessy and Evett, 2010b). Both stationary (drip irrigation systems, Osroosh et al., 2015) and moving irrigation systems (centre pivots and linear moves) lend themselves to automation and control, which to the extent of remote communication and system operation have been advanced by the irrigation industry (Kranz et al., 2012). Wireless sensors used for agricultural applications are emerging to monitor sprinkler operation and water flow, control pumps and valves, and to collect data from remote ground-based and *in situ* sensors as a method for indirectly measuring crop status (Pierce and Elliot, 2008; Vellidis et al., 2008; Mahan et al., 2010).

Hedley and Yule (2009) developed a method to predict daily spatio-temporal soil water status in EC_a defined MZs using multiple time domain transmission and time domain reflectometry soil water sensors in each MZ. These geospatial predictions are a proposed source of information to control irrigation zones on a VRI system. Kim et al. (2009) used a wireless distributed network of soil water and temperature sensors as part of a closed-loop feedback system to control irrigations on a variable rate linear move sprinkler system. Irrigations for each MZs were triggered when soil water deficit levels reached a predetermined threshold value. By means of simulations, McCarthy et al. (2014) have proposed adaptive control strategies for cotton production using model predictive control and multiple variables to synthesize dynamic prescription maps for VRI systems.

Supervisory control and data acquisition (SCADA) systems are commonly used in industrial and manufacturing applications to monitor and control processes. Key concepts of SCADA systems are monitoring, process management, adaptive control, automation, reporting and alarming (Fig. 3). System management decisions are based on information received from sensor systems monitoring processes and environments of interest. System process and control can be automated or operator-driven. System oversight and supervisory commands can be entered from a control point or remote stations to monitor and control devices (NIST, 2011).

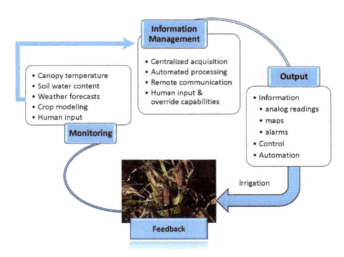

Figure 3 Pertinent conceptual points of a supervisory control and data acquisition system for production agriculture.

SCADA system concepts can be applied to production agriculture to address spatially and temporally variable crop water needs. Agricultural SCADA systems would collect data from various plant, weather and soil water sensor network systems. These data streams have traditionally been used in controlling environmental conditions of greenhouses (Sigrimis et al., 2000; Speetjens et al., 2009), and system inputs for aquaculture (Lee, 2000). In production agriculture the same data streams can help identify variable crop water needs within a field and provide precision irrigation or site-specific VRI management of fibre, grain and forage crops. The objectives of such a SCADA system would be to optimize potential yields and CWP, rather than control the environment.

Such systems generally include human to machine interfaces (HMI) through remote terminal access such as cell phones or electronic notebooks to establish set points for triggering an irrigation event, or determining how much water to apply. The HMI can also be constructed to visually indicate crop status (Andrade et al., 2015) and provide historical information such as cumulative irrigation amounts and crop yields. Also critical to the system is remote communication for monitoring the operation of the moving VRI system and maintaining integrated sensor network systems.

6 Integration of plant feedback sensor systems for site-specific VRI control

One particular type of irrigation SCADA system that is under development, is the irrigation scheduling supervisory control and data acquisition (ISSCADA) system patented by Evett et al. (2014) (Fig. 4). This system includes a client/server software programme to build prescription maps constructed from geo-referenced thermal stress indices derived from continuous weather data and canopy temperature measurements from a nearby standalone weather station and a network of infrared thermometers mounted on the VRI centre pivot pipeline. The system uses a temperature-scaling algorithm (Equation 2) to estimate diurnal temperatures using a reference temperature curve and one-time-of-day remote temperature measurements (Peters and Evett, 2004). The algorithm enables estimates of diurnal canopy temperature for each remote location that the ISSCADA system travels over in daylight hours. From these computations, temperature maps can be constructed indicating the estimated surface temperature of different areas of the field for the same time of day (Fig. 5).

$$T_{x,t} = T_0 + \frac{(T_{x'} - T_0)(T_{r,t} - T_0)}{T_{r'} - T_0} \qquad (2)$$

where $T_{x,t}$ is the estimated canopy temperature at a given location (plot) x in the field, during time interval t; $T_{r'}$ and $T_{x'}$ are, respectively, one-time-of-day canopy temperature measurements taken during the same time interval for reference location r and location x; and T_0 is the predawn temperature in the field, assumed to be uniform.

Areas southwest of the road crossing the field at 320° were fallowed, while areas northeast of the road were cropped with corn. Canopy temperatures (°F) are displayed

Deficit irrigation and site-specific irrigation scheduling techniques to minimize water use 37

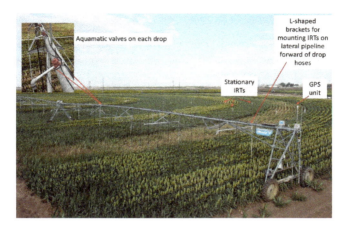

Figure 4 The early irrigation scheduling supervisory control and data acquisition (ISSCADA) system developed at Bushland, Texas, incorporates continuous canopy temperature and weather data into its decision support systems for precision irrigation management.

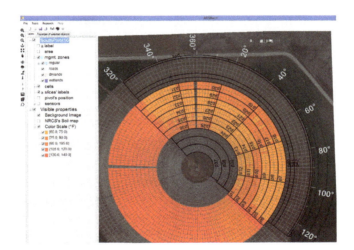

Figure 5 Map of crop canopy temperature plotted every 2° for areas of a field scanned during daylight hours of 17 August 2016, with wireless infrared thermometers (IRTs) mounted on the pipeline lateral of a variable rate irrigation centre pivot system.

with a colour scale that progresses from white (lowest temperature) to red (highest temperature). Inner- and outer-most zones did not have IRTs mounted overhead. This information could be used to indicate discreet problem areas, wherein the case of elevated temperature could point to abiotic stress such as water stress, biotic stress such as root disease or a maintenance problem with the sprinkler system such as a clogged nozzle.

Additionally, an integrated CWSI (iCWSI) based on the theoretical-based crop water stress index (CWSI) (Jackson et al., 1981) was calculated over daylight hours (Equation 3) from IRTs on a moving irrigation system and can be mapped for different areas of the field (Fig. 6).

$$\text{iCWSI}_r = \sum_{i=1}^{N} \left[\frac{(T'_c - T_a) - (T_c - T_a)_{ll}}{(T_c - T_a)_{ul} - (T_c - T_a)_{ll}} \right] \quad (3)$$

Similar to mapping surface temperature, the iCWSI can be calculated for different areas of the field for the same time of day. An irrigation SCADA system should provide recommendations for irrigation scheduling throughout the growing season. Frequent prescription maps direct site-specific irrigation according to crop water needs throughout the irrigation season, potentially improving CWP by irrigating the crop only when the crop requires water, and reducing the likelihood of excessive irrigations, run-off and deep percolation. Conversely, the systems must be robust enough to prevent significant yield loss. Automating controls or prescription maps for irrigation scheduling can help reduce a producer's time investment in managing multiple centre pivot fields.

O'Shaughnessy et al. (2015) demonstrated the use of a plant feedback ISSCADA system to adapt irrigation scheduling to a short-season cultivar of upland cotton in a semi-arid region. The dynamic prescription maps were developed for a commercial VRI centre pivot system using weather data, canopy temperature data from a moving network of wireless IRTs and canopy temperature estimations from the temperature-scaling algorithm to derive iCWSI values for MZs across the centre pivot field. Irrigation levels were varied by establishing different set points of the thermal stress index. MZs having thermal stress

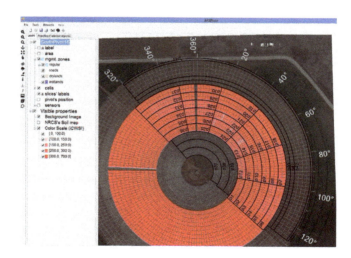

Figure 6 Map of the integrated crop water stress index (iCWSI) plotted every 2° calculated from canopy temperature scaled over daylight hours and weather data from 17 August 2016.

indexes of a higher threshold received less seasonal cumulative irrigation amounts, while MZs where the thermal stress index was lower, received greater seasonal cumulative irrigation amounts. Irrigation amounts were based on peak daily water use for the region and the frequency (days) of centre pivot travel over the entire field.

This plant feedback SCADA system provides central data collection of the canopy temperature sensor network and the standalone weather system at the base station computer located at the pivot point. The base station manages the collected data by calculating a thermal stress index for each defined MZ and builds prescription maps recommending irrigation where the established thermal stress index threshold is exceeded and no irrigation where the calculated stress index is less than the threshold. A two-year study conducted with the ISSCADA system over cotton in Bushland, Texas (O'Shaughnessy et al., 2015), resulted in water savings of 13% as compared with irrigation scheduling using weekly neutron probe readings. Lint yields (2012) were similar between irrigation methods. In the following year (2013), the irrigation amounts were similar; however, mean lint yields for the ISSCADA controlled plots were 12% greater compared with irrigation scheduling using the weekly neutron probe readings.

Other studies using SSIM have shown conservation of water and fertilizer ranging from 0 to 26% savings (Evans and King, 2012 as cited by Higgins et al., 2016). Hedley et al. (2009) described the benefits of a SSIM VRI centre pivot system in terms of water savings ranging from 9 to 19% and a reduction in deep percolation by 25–45% in SSIM compared with uniform irrigation management. King et al. (2009) did not report water savings when comparing SSIM of a VRI centre pivot to conventional uniform irrigation management. However, they reported that tuber yield distributions trended higher for SSIM areas. Although the yields were not significantly different, gross income averaged across the field was $159/ha greater for the SSIM areas.

7 Conclusions

Under scarce water conditions caused by variable seasonal rainfall, drought or limited water resources, farmers practising DI could maximize crop yields per unit of water applied more profitably than crop production with full irrigation in irrigated and rainfed regions. However, specific knowledge of crop response to DI must be incorporated in the irrigation scheduling method. Although there are risks involved, DI management schemes have been reported to reduce water use in the range of 11–54% compared with full irrigation for different crops grown in different locations around the globe. Commercialized plant and soil water sensors are available to aid in characterizing crop water stress and measuring changes of water in the soil profile. This information can be used to minimize the risk of yield loss when practising DI techniques.

Principles of site-specific irrigation explicitly include elements of precise volume, position and timing. At the core of today's high-tech systems are moving sprinklers with VRI control (speed, zone, or individual sprinkler control). These high-tech moving systems are capable of reducing water wastage by avoiding irrigation over non-arable areas and improving environmental stewardship by eliminating irrigation over water bodies and reducing the possibility of pollution from over-irrigation (9–40%), deep percolation and run-off (up to 45%). The ISSCADA system developed by ARS scientists at Bushland, Texas, is currently the most complete advanced feedback system for SSIM. Future work is required to include soil water sensing and crop growth modelling to provide a closed-loop feedback to prevent over- and under-irrigation and to improve the robustness of the decision support. Advanced technology and algorithms for irrigation scheduling and decision support are critical to facilitate sustainable agricultural production for large-size farms and for producers who manage multiple farms. As farmers continue to face challenges of limited water supplies and well capacities, precision irrigation and DI management will become critical strategies for sustaining irrigated agriculture.

8 Where to look for further information

For articles on deficit irrigation strategies, visit:

- FAO at http://www.fao.org/documents
- Proceedings papers from the Central Plains Irrigation Conference: http://www.k-state.edu/irrigate/oow/cpiadocs.html
- UC Davis Drought Management http://ucmanagedrought.ucdavis.edu/Agriculture/Crop_Irrigation_Strategies/

For more information on VRI:

- USDA-ARS, Conservation and Production Research Laboratory, Bushland, TX USA
- https://www.ars.usda.gov/people-locations/person?person-id=39705
- Annual Conferences that typically host technical session on Variable Rate Irrigation
- American Society of Agricultural & Biological Engineers (ASABE) Annual International Meeting Irrigation Association (IA) Technical Conference

9 Acknowledgements

The authors acknowledge a Cooperative Research and Development Agreement between USDA-ARS and Valmont Industries, Inc., Valley, NE (Agreement No.: 58-3K95-6-001); funding from the National Institute of Food and Agriculture, U.S. Department of Agriculture, under award number 2016-67021-24420; and support from the Ogallala Aquifer Program, a consortium between USDA-Agricultural Research Service, Kansas State University, Texas AgriLife Extension Service & Research, Texas Tech University and West Texas A&M University.

10 Disclaimer

The U.S. Department of Agriculture (USDA) prohibits discrimination in all its programmes and activities on the basis of race, colour, national origin, age, disability and where applicable, sex, marital status, familial status, parental status, religion, sexual orientation, genetic information, political beliefs, reprisal, or because all or part of an individual's income is derived from any public assistance programme. (Not all prohibited bases apply to all programmes.) Persons with disabilities who require alternative means for communication of programme information (Braille, large print, audiotape, etc.) should contact USDA's TARGET Center at (202) 720-2600 (voice and TDD). To file a complaint of discrimination, write to USDA, Director, Office of Civil Rights, 1400 Independence Avenue, S.W., Washington, D.C. 20250-9410, or call (800) 795-3272 (voice) or (202) 720-6382 (TDD). USDA is an equal opportunity provider and employer.

The mention of trade names of commercial products in this chapter is solely for the purpose of providing specific information and does not imply recommendation or endorsement by the USDA.

11 References

Acosta, J. A., Faz, A., Jansen, B., Kalbitz, K. and Martinez-Martinez, S. (2011). Assessment of salinity status in intensively cultivated soils under semiarid climate, Murcia, SE Spain. *J. Arid. Environ.* 75, 1056–66.

Agam, N., Segal, E., Peters, A., Levi, A., Dag, A., Yermiyahu, U. and Ben-Gal, A. 2014. Spatial distribution of water status in irrigated olive orchards by thermal imaging. *Prec. Agric.* 15, 346–59.

Andrade, M. A., O'Shaughnessy, S. A. and Evett, S. R. (2015). ARSmartPivot v.1- Sensor based management software for center pivot irrigation systems. Paper No. 152188736. 2015 ASABE Annual International Meeting.

Aragüés, R., Medina, E. T., Martínez-Cob, A. and Faci, J. (2014). Effects of deficit irrigation strategies on soil salinization and sodification in a semiarid drip-irrigated peach orchard. *Agric. Water Manag.* 142, 1–9.

Assefa, S., Biazin, B., Muluneh, A., Yimer, F. and Haileslassie, A. (2016). Rainwater harvesting for supplemental irrigation of onions in the southern dry lands of Ethiopia. *Agric. Water Manag.* 178, 325–34.

Aydinsakir, K., Erdal, S., Buyuktas, D., Bastug, R. and Toker, R. (2013). The influence of regular deficit irrigation applications on water use, yield, and quality components of two corn (*Zea mays* L.) genotypes. *Agric. Water Manag.* 128, 65–71.

Bainbridge, D. A. (2001). Buried clay pot irrigation: A little known but very efficient traditional method of irrigation. *Agric. Water Manag.* 48, 79–88.

Ballester, C., Castel, J., Abd El-Mageed, T. A., Castel, J. R. and Intrigliolo, D. D. (2014). Long-term response of Clementina de Nules citrus trees to summer regulated deficit irrigation. *Agric. Water Manag.* 138, 78–84.

Batchelor, C., Lovell, C. and Murata, M. (1996). Simple microirrigation techniques for improving irrigation efficiency on vegetable gardens. *Agric. Water Manag.* 32, 37–48.

Benjamin, J. G., Nielsen, D. C., Vigil, M. M., Mikha, M. M. and Calderon, F. (2015). Cumulative deficit irrigation effects on corn biomass and grain yield under two tillage systems. *Agric. Water Manag.* 159, 107–14.

Berni, J., Zarco-Tejada, P. J., Suarez, L. and Fereres, E. (2009). Thermal and narrowband multispectral remote sensing for vegetation monitoring from an unmanned aerial vehicle. *IEEE Trans. Geosci. Remote Sens.* 47(3), 722–38.

Biewald, A., Rolinski, S., Lotze-Campen, H., Schmitz, C. and Dietrich, J. P. (2014). Valuing the impact of trade on local blue water. *Ecol. Econ.* 101, 43–53.

Boelens, R. and Vos, J. (2012). The danger of naturalizing water policy concepts: Water productivity and efficiency discourses from field irrigation to virtual water trade. *Agric. Water Manag.* 108, 16–26.

Boyer, C. N., Larson, J. A., Roberts, R. K., McClure, A. T. and Typer D. D. 2014. The impact of field size and energy cost on the profitability of supplemental corn irrigation. *Agric. Sys.* 127, 61–9.

Camargo, D. C., Montoya, F., Ortega, J. F. and Córcoles, J. S. (2015). Potato yield and water use efficiency responses to irrigation in semiarid conditions. *Agron. J.* 107(6), 2120–31.

Carsky, R. J., Ndikawa, R., Singh, L. and Rao, M. R. (1995). Response of dry season sorghum to supplemental irrigation and fertilizer N and P on Vertisols in northern Cameroon. *Agric. Water Manag.* 28, 1–8.

Chai, Q., Gan, Y., Zhao, C., Xu, H., Waskom, R. M., Niu, Y. and Siddique, K. H. M. (2016). Regulated deficit irrigation for crop production under drought stress. A review. *Agron. Sustain. Dev.* 36(3), 3–21.

Chalmers, D. J., Burge, G., Jerie, P. H. and Mitchell, P. D. (1986). The mechanism of regulation of Bartlett pear fruit and vegetative growth by irrigation withholding and regulated deficit irrigation. *J. Amer. Soc. Hort. Sci.* 111, 904–7.

Chavez, M. M., Pereira, J. S., Maroco, J., Rodrigues, M. L., Ricardo, C. P. P., Osório, M. L., Carvalho, I., Faria, T. and Pinheiro, C. (2002). How plants cope with water stress in the field. Photosynthesis and growth. *Ann. Bot.* 89, 907–16.

Coates, P. W. and Delwiche, M. J. (2008). Site-specific water and chemical application by wireless valve controller network. Paper presented at the annual American Society of Agricultural and Biological Engineers, Providence.

Costello, M. J. and Patterson, W. K. (2012). Regulated deficit irrigation effect on yield and wine color of Cabernet Sauvignon in Central California. *HortSci.* 47(10), 1520–4.

Daccache, A., Knox, J. W., Weatherhead, E. K., Daneshkhah, A. and Hess, T. M. (2015). Implementing precision irrigation in a humid climate- Recent experiences and on-going challenges. *Agric. Water Manag.* 147, 135–43.

Datta, K. K. and de Jong, C. (2002). Adverse effect of waterlogging and soil salinity on crop and land productivity in northwest region of Haryana, India. *Agric. Water Manag.* 57(3), 223–38.

De la Rosa, J. M., Domingo, R., Gómez-Montiel, J. and Pérez-Pastor, A. (2015). Implementing deficit irrigation scheduling through plant water stress indicators in early nectarine trees. *Agric. Water Manag.* 152, 207–16.

Denmead, O. T. and Shaw, R. H. (1960). The effects of soil moisture stress at different stages of growth on the development and yield of corn. *Agron. J.* 52(5), 272–4.

Doorenbos, J. and Kassam, A. H. (1979). *Yield Response to Water*. FAO Irrigation and Drainage Paper No. 33, FAO, Rome, Italy, pp. 1–40.

Economic Research Service, USDA. (2013). Accessed http://www.ers.usda.gov/topics/farm-practices-management/irrigation-water-use/background.aspx (11 January 2017).

Ertek, A. and Kara, B. (2013). Yield and quality of sweet corn under deficit irrigation. *Agric. Water Manag.* 129, 138–44.

Evans, R. G. and King, B. A. (2012). Site-specific sprinkler irrigation in a water-limited future. *Trans. ASABE.* 55(2), 493–504.

Evans, R. G., LaRue, J., Stone, K. C. and King, B. A. (2013). Adoption of site-specific variable rate sprinkler irrigation systems. *Irrig. Sci.* 31, 871–87.

Fan, S. and Brzeska, J. (2016). Sustainable food security and nutrition: Demystifying conventional beliefs. *Global Food Security.* 11, 11–16.

FAO. (2002). *Deficit Irrigation Practices*. Water Reports No. 22. FAO, Rome.

FAO. (2011). *The State of the World's Land and Water Resources for Food and Agriculture (SOLAW) – Managing Systems at Risk*. Food and Agriculture Organization of the United Nations, Rome and Earthscan, London.

Fereres, E. and Soriano, M. A. (2006). Deficit irrigation for reducing agricultural water use. *J. Exp. Bot.* 58(2), 147–59.

Fracasso, A., Trindale, L. and Amaducci, S. (2016). Drought tolerance strategies highlighted by two *Sorghum bicolor* races in a dry-down experiment. *J. Plant Physiol.* 190, 1–14.

Garofalo, P. and Rinaldi, M. (2015). Leaf gas exchange and radiation use efficiency of sunflower (Helianthus annuus L.) in response to different deficit irrigation strategies: From solar radiation to plant growth analysis. *Eur. J. Agron.* 64, 88–97.

Gheysari, M., Sadeghi, S. Loescher, H. W., Amiri, S., Zareiana, M. J., Majidif, M. M., Asgariniaf, P. and Payero, J. O. (2017). Comparison of deficit irrigation management strategies on root, plant growth and biomass productivity of silage maize. *Agric. Water Manag.* 182, 126–38.

Geets, S. and Raes, D. (2009). Deficit irrigation as an on-farm strategy to maximize crop water productivity in dry areas. *Agric. Water Manag.* 96, 1275–84.

Girón, I. F., Corell, M., Martín,-Palomo, M. J., Galindo, A., Torrecillas, A. and Moreno, F. (2015). Feasibility of trunk diameter fluctuations in the scheduling of regulated deficit irrigation for table olive trees without reference trees. *Agric. Water Manag.* 161, 114–26.

Goldhamer, D. A. and Salinas, M. 2000. Evaluation of regulated deficit irrigation on mature trees grown under high evaporative demand. Proceedings Intl. Soc. Citricult. IX Congress, pp. 227–31.

Gonzalez-Dugo, V., Zarco-Tejada, P. and Nicolas, E. (2013). Using high resolution UAV thermal imagery to assess the variability in the water status of five fruit tree species within a commercial orchard. *Prec. Agric.* 14, 660–78.

Gossel, A., Thompson, A., Sudduth, K. and Henggeler, J. (2013). Performance evaluation of a center pivot variable rate irrigation system. In Proceedings: 2013 ASABE Annual International Meeting, Paper No. 131620776, Kansas City, MO, 21–24 July 2013, ASABE, St. Joseph, MI.

Grassini P., Hall. A. J. and Mercau J. L. (2009). Benchmarking sunflower water productivity in semiarid environments. *Field Crops Res.* 110, 251–62.

Grisso, R., Alley, M., Phillips, S. and McClellan, P. (2009). Interpreting Yield Maps- 'I gotta yield map – now what?'. Publication No. 442–509. College of Agriculture and Life Sciences, Virginia Polytechnic Institute and State University. http://www.pubs.ext.vt.edu/442/442-509/442-509.html

Guo, Z., Shi, Y., Yu, Z. and Zhang, Y. (2015). Supplemental irrigation affected flag leaves senescence post-anthesis and grain yield of winter wheat in the Huang-Huai-Hai Plain of China. *Field Crops Res.* 180, 100–9.

Han, Y. J., Khalili, A., Owino, T. O., Farahani, H. J. and Moore, S. (2009). Development of Clemson variable rate lateral irrigation system. *Comput. Electron. Agric.* 68, 108–13.

Hao, B., Xue, Q., Marek, T. H., Jessup, K. E., Hou, X., Xu, W., Bynum, E. D. and Bean, B. W. (2015a). Soil water extraction, water use and grain yield by drought-tolerant maize on the Texas High Plains. *Agric. Water Manag.* 155, 11–21.

Hao, B., Xue, Q., Marek, T. H., Jessup, K. E., Becker, J., Hou, X., Xu, W., Bynum, E. D., Bean, B. W., Colaizzi, P. D. and Howell, T. A. (2015). Water use and grain yield in drought-tolerant corn in the Texas High Plains. *Agron. J.* 107 (5), 1922–30.

Hatfield, J. L. and Prueger, J. H. (2010). Value of using different vegetative indices to quantify agricultural crop characteristics at different growth stages under varying management practices. *Remote. Sens.* 2(2), 562–78.

Hays, D. B., Do, J. W., Mason, R. E., Morgan, G. and Finlayson, S. C. (2007). Heat stress induced ethylene production in developing wheat grains induces kernel abortion and increased maturation in a susceptible cultivar. *Plant Sci.* 172 (6), 1113–23.

Hedley, C. B. and Yule I. J. (2009). A method for spatial prediction of daily soil water status for precise irrigation scheduling. *Agric. Water Manag.* 96,1737–45.

Hedley, C. (2014). The role of precision agriculture for improved nutrient management on farms. *J. Sci. Food Agric.*, 95, 12–19. doi: 10.1002/jsfa.6734.

Hergert, G. W., Margheim, J. F., Martin, D. L., Isbell, T. A. and Supalla, R. J. (2016). Irrigation response and water productivity of deficit to fully irrigated spring camelina. *Agric. Water Manag.* 177, 46–53.

Higgins, C. W., Kelley, J., Barr, C. and Hillyer, C. (2016). Determining the minimum management scale of a commercial variable-rate irrigation system. *Trans. ASABE.* 59(6), 1671–80.

Howell, T. A. (1990). Relationships between crop production and transpiration, evapotranspiration, and irrigation. In Steward, B.A., and Nielson, D.R., eds. Chapter 14, pp. 391–434. In *Irrigation of Agricultural Crops- Agronomy Monograph no. 30.* ASA-CSSA-SSSA, Madison, WI.

Howell, T. A. (2001). Enhancing water use efficiency in irrigated agriculture. *Agron. J.* 93(2), 281–9.

Howell, T. A., Evett, S. R., Tolk, J. A. and Schneider, A. D. (2004). Evapotranspiration of Full-, Deficit-irrigated and dryland cotton on the Northern Texas High Plains. *J. Irrig. Drain. Eng.* 130(4), 277–85.

Howell, T. A., Evett, S. R., O'Shaughnessy, S. A., Colaizzi, P. D. and Gowda, P. H. (2012). Advanced irrigation engineering: Precision and precise. *J. Agric. Sci. Tech.* A2, 1–9.

Howell, T. A., O'Shaughnessy, S. A. and Evett, S. R. (2012). Integrating multiple irrigation technologies for overall improvement in irrigation management. In Proceedings, 24th Annual Central Plains Irrigation Conference, Colby, KS, 21–22 February 2012.

Hsiao, T. C. (1973). Plant responses to water stress. *Ann. Rev. Plant Physiol.* 24, 519–70.

Hsiao, T. C. and Acevedo, E. 1974. Plant response to water deficits, water-use efficiency, and drought resistance. *Agric. Meteorol.* 14, 59–84.

Jackson, R. D., Idso, S. B., Reginato, R. J. and Pinter, P. J. (1981). Crop canopy temperature as a crop water stress indicator. *Water Resour.* 17, 1133–8.

Jones, H. G. (2004). Irrigation scheduling: Advantages and pitfalls of plant-based methods. *J. Exp. Bot.* 55(407), 2427–36.

Kang, Y., Khan, S. and Ma, X. (2009). Climate change impacts on crop yield, crop water productivity and food security- A review. *Prog. Natural Sci.* 19(12), 1665–74.

Kifle, M. and Gebretsadikan, T. G. (2016). Yield and water use efficiency of furrow irrigated potato under regulated deficit irrigation, Atsibi-Weberta, North Ethiopia. *Agric. Water Manag.* 170, 133–9.

Kim, Y., Evans, R. G. and Iversen, W. M. (2008). Remote sensing and control of an irrigation system using a wireless sensor network. *IEEE Trans. Instrument. and Measurement* 57, 1379–87.

Kim, Y., Evans, R. G. and Iversen, W. M. (2009). Evaluation of Closed-Loop Site-Specific Irrigation with Wireless Sensor Network. *J. Irrig. Drain. Engr.* 135, 25–31.

King, B. A., Stark, J. C. and Wall, R. W. (2006). Comparison of site-specific and conventional uniform irrigation management for potatoes. *Appl. Engr. ASABE.* 22(5), 677–88.

Kisekka, I., Oker, T., Nguyen, G., Aguilar, J. and Rogers, D. 2016. Mobile drip irrigation evaluation in corn. *Kansas Agric. Exp. Station Res. Rept.* 2(7), pp. 1–10. http://dx.doi.org/10.4148/2378-5977.1253.

Klocke, N. L., Currie, R. S., Tomsicek, D. J. and Koehn, J. (2011). Corn yield response to deficit irrigation. *Trans. ASABE.* 54(3), 931–40.

Koch, B., Khosla, R., Frasier, W. M., Westfall, D. and Inman, D. (2004). Economic feasibility of variable-rate nitrogen application utilizing site-specific management zones. *Agron. J.* 96(6), 1572–80.

Kranz, W. L., Evans, R. G., Lamm, F. R., O'Shaughnessy, S. A. and Peters, R. T. (2012).A review of center pivot irrigation control and automation technologies. *Appl. Eng. Agric.* 28(3), 389–97.

Kurian, T. 1975. Effect of supplemental irrigation with seas water on growth and chemical composition of Pearl Millet (Pennisetum typhoides S.etH.). *Z. Pflanzenphysiol.* 79(5), 377–83.

Kusakabe, A., Contreras-Barragean, B. A., Simpson, C. R., Enciso, J. M., Nelson, S. D. and Melgar, J. C. (2016). Application of partial rootzone drying to improve irrigation water use efficiency in grapefruit trees. *Agric. Water Manag.* 178,: 66–75.

Lamm, F. R., Aiken, R. M. and Kheira, A. A. (2009). Corn yield and water use characteristics as affected by tillage, plant density and irrigation. *Trans. ASABE.* 52(1), 133–43.

Lanari, V., Palliotti, A., Sabbatini, P. and Howell, G. S. (2014). Optimizing deficit irrigation strategies to manage vine performance and fruit composition of field-grown 'Sangiovese' (Vitis vinifera L.). *Sci. Horticult.* 179, 239–47.

LaRue, J. and Frederick, C. (2012). Decision process for the application of variable rate irrigation. In Proceedings: 2012 ASABE Annual International Meeting. Paper No. 12–1337451, Dallas, TX, 29 July–1 August 2012, ASABE, St. Joseph, MI.

Liu, C., Rubaek, G. H., Liu, F. and Andersen, M. N. (2015). Effect of partial root zone drying and deficit irrigation on nitrogen and phosphorus uptake in potato. *Agric. Water Manag.* 159, 66–76.

Liu, J., Sun, S., Wu, P., Wang, Y. and Zhao, X. (2015). Inter-county virtual water flows of the Hetao irrigation district, China: A new perspective for water scarcity. *J. Arid Environ.* 119, 31–40.

Lokhande, S. and Reddy, K. R. (2014). Reproductive and fiber quality responses of upland cotton to moisture deficiency. *Agron. J.* 106, 1060–9.

Lyle, W. M. and Bordovsky, J. P. (1983). LEPA irrigation system evaluation. *Trans. ASAE* 26(3), 776–781.

Ma, L., Ahuja, L. R., Islam, A., Trout, T. J., Saseendran, S. A. and Malone, R. W. (2017). Modelling yield and biomass responses of maize cultivars to climate change under full and deficit irrigation. *Agric. Water Manag.* 180, 88–98.

Madramootoo, C. A. and Jutras, P. J. (1984). Supplemental irrigation of bananas in St. Lucia. *Agric. Water Manag.* 9, 149–56.

Mahan, J. R., Conaty, W., Neilson, J., Payton, P. and Cox, S. B. (2010). Field performance in agricultural settings of a wireless temperature monitoring system based on a low-cost infrared sensor. *Comput. Electron. Agric.* 71, 176–81.

McCarthy, A. C., Hancock, N. H. and Raine, S. R. (2010). VARIwise: A general-purpose adaptive control simulation framework for spatially and temporally varied irrigation at sub-field scale. *Comput. Electron. Agric.* 70, 117–28.

McCarthy, A. C., Hancock, N. H. and Raine, S. R. (2014). Simulation of irrigation control strategies for cotton using model predictive control within the VARIwise simulation framework. *Comput. Electron. Agric.* 101, 135–47.

Milton, A. W., Perry, C. D. and Khalilian, A. (2006). Status of Variable-rate irrigation in the southeast. In Proceedings: 2006 ASABE Annual International Meeting, Paper No. 061075, 9–12 July 2006, ASABE, St. Joseph, MI.

Mounce, R. B., O'Shaughnessy, S. A., Blaser, B. C., Colaizzi, P. D. and Evett, S. R. 2016. Crop response of drought-tolerant and conventional maize hybrids in a semiarid environment. *Irrig. Sci.* 34(30), 231–44.

Mounzer, O., Pedrero-Salcedo, F., Nortes, P. A., Bayona, J., Nicolás, E. and Alarcón, J. J. (2013). Transient soil salinity under the combined effect of reclaimed water and regulated deficit drip irrigation of Mandarin trees. *Agric. Water Manag.* 120, 23–9.

National Agricultural Statistic Service (NASS), USDA, Census of Agriculture. (2013). 2012 Census of Agriculture- Farm and Ranch Irrigation Survey (2013). https://www.agcensus.usda.gov/Publications/2012/Online_Resources/Farm_and_Ranch_Irrigation_Survey/ (accessed 26 December 2016).

NIST (National Institute of Standards and Technology). (2011). *Guide to Supervisory Control and Data Acquisition (SCADA) and Industrial Control Systems Security* (Eds: Stouffer, K., Falco, J. and Kent, K.). Special Publication 800–82. U.S. Department of Commerce. http://nvlpubs.nist.gov/nistpubs/SpecialPublications/NIST.SP.800-82.pdf

Olson, B. L. S. and Rogers D. H. (2008). Comparing drag hoses verses sprinklers on corn irrigated by a center pivot. *Appl. Engr. ASABE.* 24(1), 41–5.

O'Shaughnessy, S. A. and Evett, S. R. (2010a). Canopy temperature based system effectively schedules and controls center pivot irrigation of cotton. *Agric. Water Manage.* 97, 1310–16.

O'Shaughnessy, S. A. and Evett, S. R. (2010b). Developing wireless sensor networks for monitoring crop canopy temperature using a moving sprinkler system as a platform. *Appl. Engr. Agric.* 26, 331–41.

O'Shaughnessy, S. A., Urrego, Y. F., Evett, S. R., Colaizzi, P. D. and Howell, T. A. (2013). Assessing application uniformity of a variable rate irrigation system in a windy location. *Appl. Engr. Agric.* 29(4), 497–510.

O'Shaughnessy, S. A., Evett, S. R., Colaizzi, P. D., Tolk, J. A. and Howell T. A. (2014). Early and late maturing grain sorghum under variable climatic conditions in the Texas High Plains. *Trans. ASABE.* 57(6), 1583–894.

O'Shaughnessy, S. A., Evett, S. R. and Colaizzi, P. D. (2015). Dynamic prescription maps for site-specific variable rate irrigation of cotton. *Agric. Water Manage.* 159, 123–38.

O'Shaughnessy, S. A., Evett, S. R., Andrade, A., Workneh, F., Price, J. A. and Rush, C. M. (2016). Site-Specific Variable Rate Irrigation as a Means to Enhance Water Use Efficiency. *Trans. ASABE* 59(1), 239–49. doi: 10.13031/trans.59.11165.

Osroosh, Y., Peters, R. T., Campbell, C. S. and Zhang, Q. (2015). Automatic irrigation scheduling of apple trees using theoretical crop water stress index with an innovative dynamic threshold. *Comput. Electron. Agric.* 118, 193–203.

Oweis, T. (1997). *Supplemental Irrigation: A Highly Efficient Water-use Practice.* ICARDA: Aleppo, Syria, 16.

Oweis, T. and Hachum, A. (2009). Optimizing supplemental irrigation: Tradeoffs between profitability and sustainability. *Agric. Water Manage.* 96, 511–16.

Payero, J. O., Melvin S. R., Irmak, S. and Tarkalson, D. (2006). Yield response of corn to deficit irrigation in a semiarid climate. *Agric. Water Manag.* 84, 101–12.

Peters, R. T. and Evett, S. R. (2004). Modeling diurnal canopy temperature dynamics using one-time-of-day measurements and a reference temperature curve. *Agron. J.* 96, 1553–61.

Peters, T. R., and Evett, S. R. (2008). Automation of a Center Pivot using the temperature-time threshold method of irrigation scheduling. *J. Irrig. Drain. Eng.* 134, 286–91.

Phene, C. J., Howell, T. A. and Sikorski, M. D. (1985). A traveling trickle irrigation system. In Hillel, D. (Ed.) *Advances in Irrigation* (pp. 1–47). London: Academic Press.

Pláyan, E. and Mateos, L. (2006). Modernization and optimization of irrigation systems to increase water productivity. *Agric. Water Manag.* 80, 100–16.

Rawlins, S. L., Hoffman. G. W. and Merrill, S. D. (1974). Traveling trickle system. In Proceedings of Int. Drip Irrig. Congr., 2nd, San Diego, pp. 184–7.

Rey, D., Holman, I. P., Daccache, A., Morris, J., Weatherhead, E. K. and Knox, J. W. (2016). Modelling and mapping the economic value of supplemental irrigation in a humid climate. *Agric. Water Manag.* 173: 13–22.

Robins, J. S. and Domingo, C. E. (1953). Some effects of severe soil moisture deficits at specific growth stages in corn. *Agron. J.* 45(12), 618–21.

Robins, J. S. and Domingo, C. E. (1956a). Moisture deficits in relation to the growth and development of dry beans. *Agron. J.* 48(2), 67–70.

Robins, J. S. and Domingo, C. E. (1956b). Potato yield and tuber shape as affected by severe soil-moisture deficits and plant spacing. *Agron J.* 48(11), 488–92.

Robins, J. S. and Domingo, C. E. (1962). Moisture and nitrogen effects on irrigated spring wheat. *Agron J.* 54(2), 135–8.

Romero, P., Gil-Munoz, R., del Amor, F. M., Valdes, E., Fernandez, J. I. and Martinez-Cutillas, A. (2013). Regulated deficit irrigation based upon optimum water status improves phenolic composition in Monastrell grape and winds. *Agric. Water Manag.* 121, 85–101.

Rosenberg, O., Cohen, Y., Saranga, Y., Levi, A. and Alchanatis, V. (2013). Comparison of methods for field scale mapping of plant water status using aerial thermal imagery, pp. 185–92. In Stafford, J. (Ed.), *Precision Agriculture '13*. Wageningen Academic Publishers. https://www.researchgate.net/profile/Y_Cohen/publication/265297616_Comparison_of_methods_for_field_scale_mapping_of_plant_water_status_using_aerial_thermal_imagery/links/54081b4f0cf23d9765ae3e41.pdf

Roth, J. A, Ciampitti, I. A. and Vyn, T. J. (2013). Physiological evaluations of recent drought-tolerant maize hybrids at varying stress levels. *Agron. J.* 105(4), 1129–41.

Sadler, E. J., Evans, R. G., Stone, K. C. and Camp, C. R. (2005). Opportunities for conservation with precision irrigation. *J. Soil Water Conser.* 60, 371–9.

Sampathkumar, T., Pandian, B. J., Rangaswamy, M. V., Manickasundaram, P. and Jeyakumar, P. (2013). Influence of deficit irrigation on growth, yield and yield parameters of cotton-maize cropping sequence. *Agric. Water Manag.* 130, 90–102.

Setia, R., Marschner, P. Baldock, J., Chittleborough, D., Smith, P. and Smith, J. (2011). Salinity effects on carbon mineralization in soils of varying texture. *Soil Biol. Biochem.* 43, 1908–16.

Shackel, K. (2011). A plant-based approach to deficit irrigation in trees and vines. *HortSci.* 46(2), 173–7.

Sharp, R. E. and Davies, W. J. (1979). Solute regulation and growth by roots and shoots of water stressed maize plants. *Planta.* 147,43–9.

Sheng Han, S. (1974). *Fan Sheng-chih Shu: An Agriculturist Book of China written by Fan sheng-chih in the First Century BC*. Science Books, Peking, pp. 36–7.

Sinclair, T. (2011). Challenges in breeding for yield increase for drought. *Trends in Plant Sci.* 16(6), 289–93.

Speetjens, S. L., Stigter, J. D. and van Straten, G. 2009. Towards an adaptive model for greenhouse control. *Comput. Electron. Agric.* 67(1–2), 1–8.

Spreer, W., Nagle, M., Neidhart, S., Carle, R., Ongprasert, S. and Müller, J. (2007). Effect of regulated deficit irrigation and partial rootzone drying on the quality of mango fruits. *Agric. Water Manag.* 88, 173–80.

Stone, K. C., Bauer, P. J. and Sigua, G. C. (2016). Irrigation management using an expert system, soil water potentials, and vegetative indices for spatial applications. *Trans. ASABE*. 59(3), 941–8.

Tari, A. F. (2016). The effects of different deficit irrigation strategies on yield, quality, and water-use efficiencies of wheat under semi-arid conditions. *Agric. Water Manag*. 167, 1–10.

Tarjuelo, J. M., Rodriguez-Diaz, J. A., Abadía, R., Camacho, E., Rocamora, C. and Moreno M. A. (2015). Efficient water and energy use in irrigation modernization: Lessons from Spanish case studies. *Agric. Water Manag*. 162, 67–77.

Thorpe, K. R., Hunsaker, D. J., French, A. N., Bautista, E. and Bronson, K. F. (2015). Integrating geospatial data and cropping system simulation within a geographic information system to analyze spatial seed cotton yield, water use, and irrigation requirements. *Prec. Agric*. 16, 532–57.

Tolk, J. A. and Howell, T. A. (2012). Sunflower water productivity in four Great Plains soils. *Field Crops Res*. 127, 120–8.

Vadez, V. (2014). Root hydraulics: The forgotten side of roots in drought adaptation. *Field Crops Res*. 165, 15–24.

Vellidis, G., Tucker, M., Perry, C., Kvien, C. and Bednarz, C. (2008). A real-time wireless smart sensor array for scheduling irrigation. *Comput. Electron. Agric*. 61,44–50.

Vories, E. D., Stevens, W., Rhine, M. and Straatmann, Z. (2016). Investigating irrigation scheduling for rice using variable rate irrigation. *Agric. Water Manag*. 179, 314–23.

Wang, D., Yu, Z. and White, P. J. (2013). The effect of supplemental irrigation after jointing on leaf senescence and grain filling in wheat. *Field Crops Res*. 151, 35–44.

Wang, Y. S., Liu, F. L., Jensen, L. S., de Neergaard, A. and Jensen, C. R. (2013). Alternate partial root-zone irrigation improves fertilizer-N use efficiency in tomatoes. *Irrig. Sci*. 31, 589–98.

Wilchelns, D. and Qadir, M. (2015). Achieving sustainable irrigation requires effective management of salts, soil salinity, and shallow ground water. *Agric. Water Manag*. 157, 31–8.

Yan, N., Marschner, P. Cao, W., Zuo, C. and Qin, W. (2015). Influence of salinity and water content on soil microorganisms. *Int. Soil Water Conserv. Res*. 3, 316–23.

Zhang, H. and Wang, D. (2013). Management of postharvest deficit irrigation of peach trees using infrared canopy temperature. *Vadose Zone J*. 12(3), 1–11. doi:10.2136/vzj2012.0093.

Zude-Sasse, M., Fountas, S., Gemtos, T. A. and Abu-Khalaf, N. (2016). Applications of precision agriculture in horticultural crops. *Eur. J. Hortic. Sci*. 81(2), 78–90.

Chapter 3

Improving water management in winter wheat

Q. Xue, J. Rudd, J. Bell, T. Marek and S. Liu, Texas A&M AgriLife Research and Extension Center at Amarillo, USA

1 Introduction
2 Winter wheat yield
3 Yield determination under water-limited conditions
4 The role of measuring evapotranspiration (ET)
5 Water-use efficiency
6 Wheat yield, evapotranspiration (ET) and water-use efficiency (WUE) relationships
7 Case studies
8 Future trends and conclusion
9 Where to look for further information
10 References

1 Introduction

Wheat (*Triticum aestivum* L. em. Thell) is a major crop in the world and is the number one food grain consumed directly by humans. Wheat is grown in a wide range of environments around the world and has the broadest adaptation of all cereal crop species (Briggle and Curtis, 1987). Therefore, wheat production plays a critical role in the world economy and food security. For 2016/17, the world wheat production is projected to be about 720 million metric tons. The European Union dominates the production with 157 million tons, followed by China (130 million tons), India (88 million tons), Russia (64 million tons), the United States (56 million tons), Canada (28.5 million tons), Pakistan (25.3 million tons), Australia (25 million tons), Ukraine (24 million tons) and Turkey (17.5 million tons) (http://www.statista.com/statistics/237912/global-top-wheat-producing-countries/). Wheat is also widely traded in the world market, with the United States, European Union, Australia, Canada and Russia being the major exporters (Fischer et al., 2014). For example, the United States share of world wheat exports has ranged from about 16% to nearly 40% in last three decades (Fig. 1). Although wheat exports have a declining trend in coming decade, the United States will remain one of the world's leading suppliers of high-quality

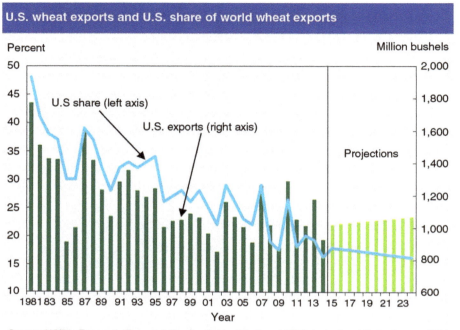

Source: USDA, Economic Research Service using data from *USDA Agricultural Projections to 2024*, February 2015.

Figure 1 U.S. wheat exports and share of world wheat exports (historic data and projections to 2024).

wheat (USDA-Economic Research Service, 2015) (http://www.ers.usda.gov/data-products/chart-gallery/detail.aspx?chartId=52238).

Globally, winter wheat is mainly grown in Eurasia, North China Plain (NCP), Anatolia area of Iran and the United States. Some areas have high precipitation (>500 mm) such as Europe and eastern United States but other areas such as NCP require irrigation for high yield in winter wheat (Fischer et al., 2014). Nevertheless, drought stress can significantly reduce winter wheat yields, even under high precipitation environment like United Kingdom (Foulkes et al., 2002). In the United States, except the eastern states, winter wheat is mainly produced in the Great Plains and Pacific Northwest states, with a low to moderate precipitation (200–500 mm). The impact of drought stress on winter wheat production in these areas is more significant than others. The Great Plains of North America extend from the northern part of the state of Coahuila, Mexico, through west Texas and northward through the United States to the prairie provinces of Alberta, Manitoba and Saskatchewan of Canada (Fig. 2). In the United States, central and southern Great Plains (SGP) are major winter wheat production areas. In 2015, winter wheat production from the states of Colorado, Kansas, New Mexico, Oklahoma and Texas was 665 million bushels (18 million metric tons) and accounted for 42% of US total winter wheat production (Fig. 3; NASS, 2016).

The SGP includes the Texas High Plains, the Oklahoma panhandle, part of eastern New Mexico, southwestern Kansas and southeastern Colorado. Winter wheat is widely

Figure 2 Great Plains of North America.

grown under dryland (rainfed), full irrigation and deficit irrigation production systems and produced for both grain and winter cattle forage in this region (Musick and Dusek, 1980; Musick et al., 1994; Howell et al., 1995). The region has a semi-arid climate with annual precipitation ranging from 380 mm in the southwest to 580 mm in the northeast and averages about 480 mm. Growing season precipitation for wheat production averages about 250 mm (Musick et al. 1994). The seasonal evapotranspiration (ET) for winter wheat growth ranges from 700 to 950 mm under full irrigation conditions (Musick and Porter, 1990; Musick et al., 1994; Howell et al. 1995, 2007). Therefore, the seasonal precipitation for winter wheat can only meet one-third of the ET required for maximum grain yield.

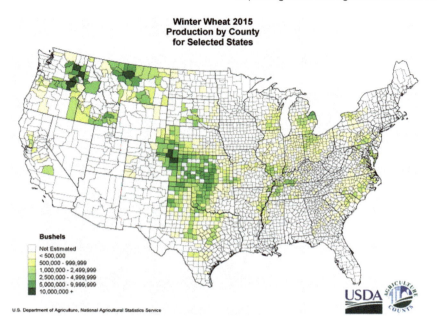

Figure 3 Winter wheat production by county in the United States in 2015 (NASS, 2016). https://www.nass.usda.gov/Charts_and_Maps/Crops_County/ww-pr.php.

As a result, wheat yield and water-use efficiency (WUE) are primarily limited by soil water deficits from late spring to early summer (Musick et al., 1994; Howell et al., 1997). Under dryland conditions, wheat production is largely determined by the amount and effective use of soil water storage and seasonal precipitation. In general, dryland wheat yields are much lower than irrigated wheat in the region due to the drought stress (Musick et al., 1994; Jones and Popham, 1997).

In the U.S. SGP, high wheat yields have been dependent on irrigation. The irrigation water resource is mainly from the Ogallala Aquifer (Fig. 4), and development of irrigation in this region significantly increased during the 1950s. The Ogallala Aquifer is essentially a closed system with minimal recharge capacity, and the dramatic increase in water extraction for crop irrigation resulted in a significant decline in the water table; some areas have experienced up to 50% reduction in predevelopment saturated thickness. Irrigated land area has decreased from a peak of 2.4 million ha in 1974 to 1.9 million ha in 2000 (Colaizzi et al., 2008).

Similar to many other agricultural regions in the world, the SGP faces many challenges in crop production. First, the growing world population continuously requires more production of food, forage and fibre, particularly the major food crops such as wheat and corn. Second, the declining water table in the Ogallala Aquifer and increasing pumping costs will inevitably reduce irrigation levels (Musick et al. 1994; Stone and Schlegel, 2006; Colaizzi et al., 2008). Third, the possibility of increasing frequency and severity of drought stress as well as other abiotic and biotic stresses under changing climate will likely reduce crop yields more frequently. For example, the wheat yield and economic losses from the

Figure 4 Ogallala aquifer.

historic 2011 drought in Texas resulted in $5.2 billion agricultural losses and the losses from wheat alone were $243 million. The last two drought years in 2006 and 2009 resulted in $4.1 and $3.6 billion economic losses in Texas agriculture, respectively (Fannin, 2011).

NCP produces about 69% of wheat in the whole country (Mo et al., 2009). The average annual rainfall in the region ranges from 500 to 650 mm. However, the majority of precipitation occurs during summer months (July–September). Precipitation during the wheat growing season ranges from 100 to 180 mm, or approximately 25–40% of crop water requirement over the growing season (Zhang et al., 1999). Therefore, drought is also an important factor affecting wheat yield without irrigation. Similar to the U.S. SGP, irrigation from groundwater sources is also mainly used in winter wheat production. However, groundwater levels have also decreased considerably in the NCP due to irrigation (Zhang et al., 2003).

Since water is the most important factor affecting crop production, development of crop management practices to conserve water, optimize water use and improve crop WUE becomes essential, particularly under changing climate conditions. Although increasing yield is an ultimate goal to any crop production system, maximizing WUE is of particular importance under water-limited conditions since lower WUE sometime indicates poor management and inefficient use of available soil water (Musick et al., 1994; Howell, 2001; Passioura and Angus, 2010; Xue et al., 2012).

In this chapter, we reviewed the research progress in winter wheat water management and WUE. The major data sources were from long-term field experiments in the U.S. SGP due to the long history of winter wheat research. We also include some research results from NCP of China as irrigation studies have been conducted extensively in the past two decades. We initially discuss the yield, ET, WUE and their relationships. Then, the best management practices will be discussed based on several case studies, ranging from soil and water conservation to genetic improvement of drought tolerance to deficit irrigation practice.

2 Winter wheat yield

Winter grain yield is determined by genetic, environmental and management factors. In the U.S. SGP, water is the major limiting factor for wheat yield. The long-term dryland wheat yields at Bushland, Texas (1984–93), were mostly in range of 1–2 Mg ha^{-1} (Jones and Popham, 1997). Musick and Porter (1990) reported that farm level dryland yield averaged 1.1 Mg ha^{-1} from 1968 to 1986 in the SGP. However, yields in dryland treatment from irrigation studies were frequently over 3 Mg ha^{-1} (Musick et al., 1994; Xue et al., 2006). Stone and Schlegel (2006) summarized long-term winter wheat yield in Tribune, Kansas (1973–2004), and reported an average yield of 2.5 Mg ha^{-1}. Graybosch and Peterson (2010) analysed winter wheat yield in the Great Plains based on the yield data from USDA-coordinated winter wheat regional performance nurseries from 1959 to 2008. They showed that dryland wheat yield ranged from 1.8 to 2.9 Mg ha^{-1}, with an average of 2.3 Mg ha^{-1} in the SGP (Graybosch and Peterson, 2010). This average yield is close to the average yield from Stone and Schlegel (2006). Genetic improvement in wheat yield was evident over the years based on the long-term yield data, with 1.3% yield increase per year up to the middle of 1990s (Graybosch and Peterson, 2010). Our recent studies also demonstrated a significant genetic gain in dryland wheat yield at Bushland, Texas. For example, in a two-year field study, relatively newer cultivar TAM 111 (released in 2003) yielded 2.8 Mg ha^{-1} but a relatively older cultivar, TAM 105 (released in 1979), only yielded 2.3 Mg ha^{-1} in 2010 (Xue et al., 2014).

In the SGP, irrigated wheat yields ranged from 3.0 to 7.7 Mg ha^{-1}, depending on irrigation timing and frequency, and environmental conditions (Xue et al., 2012). In general, yield increased as irrigation amount and frequency increased (Schneider and Howell, 2001; Xue et al., 2006). Under full irrigation, yield genetic gains are more significant as compared to dryland. At Bushland, Texas, long-term irrigated studies showed that the fully irrigated yield in 1980s was in the range of 4–5 Mg ha^{-1} (Musick and Dusek, 1980) but increased to about 7.3 Mg ha^{-1} in the late 1990s (Schneider and Howell, 2001). The high yield of 7.7 Mg ha^{-1} was also found in producer's research and demonstration fields (AgriPartner, 2007). The highest irrigated yield in Tribune, Kansas, was about 6.0 Mg ha^{-1}, which was likely under limited irrigation conditions (Stone and Schegel, 2006). In NCP, winter wheat is mainly grown under irrigation and yields have increased over the past four decades. For example, Zhang et al. (2010) showed that cultivars released in 2000s yielded about 7.0 Mg ha^{-1} as compared to yield of 5.0 Mg ha^{-1} in 1970s. The more recent field studies indicated that irrigated yield can be over 8.0 Mg ha^{-1} (Wang et al., 2016).

3 Yield determination under water-limited conditions

For yield determination under water-limited conditions, several conceptual models have been introduced by different authors (Passioura, 1977; Howell, 2001; Richards et al., 2002; Blum, 2009; Passioura and Angus, 2010; Stewart and Peterson, 2015). For wheat, Passioura (1977) introduced a framework for identifying the important components for yield under water-limited conditions. Grain yield is determined by three components:

$$\text{Yield} = ET \times WUE_{bm} \times HI \qquad (1)$$

where ET is the seasonal evapotranspiration, WUE_{bm} is the water-use efficiency for biomass production, and HI is the harvest index, that is, the fraction of biomass partitioning to grains (Passioura, 1977). Since these three components are likely to be largely independent of each other, then an improvement in any one of them should result in an increase in yield. This framework has been proved to be very useful for identifying management strategies under water-limited conditions (Richards, et al., 2002).

Stewart and Peterson (2015) introduced another form of yield determination as

$$\text{Yield} = ET \times T/ET \times 1/TR \times HI \qquad (2)$$

where T/ET is the ratio of plant transpiration and evapotranspiration and TR is the transpiration ratio (the amount of water required to produce a unit of biomass). In equation (1), WUE_{bm} is related to transpiration efficiency (TE), transpiration (T) and soil evaporation (Es) (Richards et al., 2002):

$$WUE_{bm} = TE/(1+ Es/T) \qquad (3)$$

In equation (2), 1/TR is actually transpiration efficiency (TE).

Based on the above equations, increasing ET and HI and minimizing Es will increase yield under water-limited conditions. In general, TE is a relatively conservative parameter due to the low variability in the basic biochemical efficiency of photosynthesis (Bodner et al., 2015).

In equations (1) and (2), the product of ET × WUE_{bm} or ET × T/ET × 1/TR is biomass at maturity (BM). Therefore, the yield determination equation also can be simply rewritten as follows:

$$\text{Yield} = BM \times HI \qquad (4)$$

Improving biomass production, HI or both will lead to higher yield (Blum, 2009). In the southern Great Plains, an early analysis indicated that wheat improvement from 1950s to 1980s was mainly contributed by increased biomass. The potential HI was relatively stable in the semi-dwarf cultivars (Howell, 1990). In recent studies, we also found that higher yield in newer cultivars was a result of higher biomass under dryland conditions

(Xue et al., 2014). The HI has increased by genetic improvement as evidenced in our new study. However, HI is generally low due to drought or heat stress at reproductive stages in semi-arid environments (Schneider and Howell, 2001).

4 The role of measuring evapotranspiration (ET)

ET has been widely used in determining water requirements, irrigation scheduling and assessing crop growth and yield response to water deficit. In the U.S. SGP, seasonal ET for winter wheat has been measured in irrigation studies since 1960s. Musick and Porter (1990) summarized ET requirement in wheat up to late 1980s. Xue et al. (2012) updated the information to early 2000s in the region. Under full irrigation conditions, seasonal ET for winter wheat averaged 710 mm in Bushland, Texas (Musick and Porter, 1990). Xue et al. (2006) reported a similar seasonal ET (691 mm) in a field study in 1993. However, Howell et al. (1995) showed that the seasonal ET under full irrigation was up to 957 mm in a 3-year lysimeter study. Schneider and Howell (2001) showed that seasonal ET at 100% ET requirement averaged 825 mm under centre pivot system. In the NCP, seasonal ET of winter wheat is generally in the range of 400–500 mm, much lower than the U.S. SGP (Hao et al., 2014).

5 Water-use efficiency

In most field studies, WUE is simply defined as the ratio of grain yield and seasonal ET and expressed as kg m^{-3}. Another common unit of WUE is kg ha^{-1} mm^{-1}. Since WUE is the ratio of yield and ET, either higher yield or lower ET can result in higher WUE. However, increasing yield is an ultimate goal for crop production. As such, a higher WUE value should reflect an increased yield, not reduced yield. Economic benefits from increased WUE under water-limited conditions are usually achieved only if yield is maximized for the available water (Sinclair and Muchow, 2001). Frequently, a lower WUE value could be the result of poor management or adverse environmental conditions such as water stress (Musick et al., 1994). In the U.S. SGP, WUE ranged from 0 to 0.8 kg m^{-3}, with an average of 0.4 kg m^{-3} for dryland wheat based on long-term field studies at Bushland, Texas (Jones and Popham, 1997). The major reason for such low WUE values is because of drought stress in dryland wheat in the SGP (Musick and Porter, 1990). In contrast, WUE in irrigated wheat was much higher than that in dryland and ranged from 0.5 to 1.2 kg m^{-3} (Musick et al., 1994; Xue et al., 2006). Therefore, irrigation management plays an important role to increase wheat yield and WUE in the SGP. Since wheat is a relatively drought tolerant crop as compared to corn, deficit irrigation can significantly increase WUE (Musick et al., 1994). In the NCP, WUE in winter wheat is generally higher than that in the U.S. SGP. Hao et al. (2014) summarized the winter wheat irrigation studies in NCP and showed that WUE ranged from 0.9 to 2.3 kg m^{-3}, with an average about 1.6 kg m^{-3}. The main reason for the higher WUE in NCP is high yield (up to 9 Mg ha^{-1}) but relatively low seasonal ET (generally less than 600 mm).

Over the years, WUE has been improved through breeding efforts as evidenced in both the SGP and the NCP. In SGP, Musick and Porter (1990) showed that wheat WUE increased from 0.44 kg m^{-3} with the cultivar Concho grown in 1950s to 0.54 kg m^{-3} with the cultivar Tascosa in the late 1960s and to 0.94 kg m^{-3} for the semi-dwarf cultivar in the early 1980s. Schneider and Howell (2001) reported a higher WUE of 1.14 kg m^{-3} in late 1990s. In the NCP, Zhang et al. (2010) demonstrated wheat WUE increased from 1.0 to 1.2 kg m^{-3} for cultivars from the early 1970s to 1.4–1.5 kg m^{-3} for recently released cultivars in Northern China Plain.

6 Wheat yield, evapotranspiration (ET) and water-use efficiency (WUE) relationships

The wheat yield-ET relationship has been reported in different studies (Musick et al., 1994; Schneider and Howell, 2001; Stone and Schlegel, 2006; Xue et al., 2012; Hao et al., 2014). Musick et al. (1994) summarized the yield-ET relationship based on long-term data in dryland and irrigated plots from 1958 to 1992 at Bushland, Texas. They showed a linear relationship between grain yield and seasonal ET pooling dryland and irrigated data together. The linear regression between yield and ET resulted in a slope of 1.22 kg grain yield per m^3 of seasonal ET (kg m^{-3}) and a threshold of 206 mm ET (Musick et al., 1994). Xue et al. (2012) analysed the yield-ET relationship again by updating the data to late 1990s. Similarly, there was a significant linear relationship between yield and ET, and the regression resulted in a slope of 1.06 kg m^{-3} and a threshold of 164 mm ET ($Y = 0.0106X-1.7393$, $R^2 = 0.75$, $P < 0.001$) (Fig. 5). The new data showed a slightly lower slope but a lower ET threshold as compared to Musick et al. (1994). Stone and Schlegel (2006) summarized dryland wheat data from 1974 to 2004 and showed a linear relationship between wheat yield and ET ($Y = 0.01X-1.838$, $R^2 = 0.64$, $P < 0.0001$). The slope was 1.0 kg

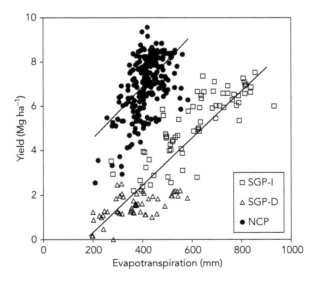

Figure 5 The linear relationship between yield and evapotranspiration (ET) in the U.S. SGP and NCP. I–irrigated; D–dryland. Regression equations: SHP-$Y = 0.0106X-1.7393$, $R^2 = 0.75$, $P < 0.001$; NCP-$Y = 0.0117X + 2.1945$, $R^2 = 0.31$, $P < 0.001$. Sources: Xue et al. (2012); Hao et al. (2014).

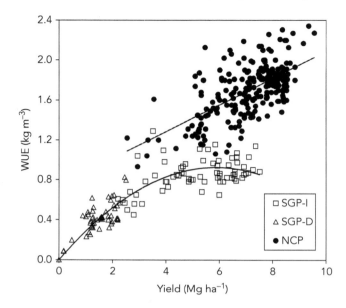

Figure 6 The relationship between WUE and yield in the U.S. SGP and NCP. I–irrigated; D–dryland. Regression equations: SGP-$Y = -0.0267X^2+0.3138X$, $R^2=0.79$, $P < 0.001$; NCP-$Y = 0.1338X + 0.7482$, $R^2 = 0.40$, $P < 0.001$). (Sources: see Fig. 4).

m^{-3} and the ET threshold was 183 mm, which were close to the results from Bushland in Fig. 5. The WUE based on the above linear regression analysis was about 1.0 kg m^{-3} in the SGP region.

Hao et al. (2014) summarized field studies in multiple years in NCP. There was also a linear relationship between yield and seasonal ET ($Y = 0.0117X + 2.1945$, $R^2 = 0.31$, $P < 0.001$) (Fig. 5). The linear regression resulted in a slope of 1.17 kg m^{-3}. However, the ET threshold could not be estimated because of the more scattered data and low R^2 value. The more scattered data in NCP was because of the multi-locations while data in SGP were mainly from one or two locations (Bushland, TX, or Tribune, KS). Nevertheless, the linear regression of yield and ET in NCP had a greater slope (1.17 kg m^{-3}) in NCP than that in SHP (1.06 kg m^{-3}).

Figure 6 showed the relationship between WUE and wheat yield in both SGP and NCP. The combined data showed a clear linear relationship between WUE and yield ($Y = 0.2086X + 0.1175$, $R^2 = 0.75$, $P < 0.0001$). However, the WUE-yield relationship was a quadratic function when the full range of yield was considered for SGP. The WUE increased linearly when yield increased up to 4–5 Mg ha^{-1}. When yield increased further, WUE maximized and even tended to decrease. In NCP, the WUE increased linearly as yield increased. Fig. 6 indicates that higher WUE generally can be achieved with higher yields. However, curvilinear relationship between WUE and yield in the SHP showed that WUE might not be the highest when yield was in the high range. The ET demand could be as high as 12 mm per day and was frequently over 60 mm per week in irrigated wheat due to high winds and associated high vapour pressure deficit (Howell et al., 1995) (Fig. 7). In the NCP, daily ET was generally less than 10 mm (Liu et al., 2002).

Improving water management in winter wheat 59

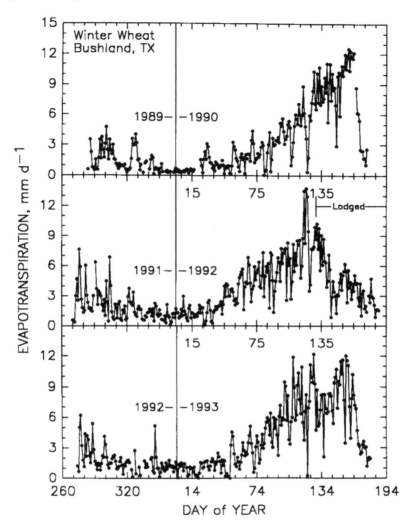

Figure 7 Daily evapotranspiration during winter wheat growing season under fully irrigated conditions in Bushland, Texas (Howell et al., 1995, used with permission).

7 Case studies

The above analysis of yield, ET, WUE and their relationships demonstrates that there are large variations in yield and WUE at different ET levels for winter wheat production. As such, development of management practices is important to improve yield and WUE under water-limited conditions. Generally, there are two ways to improve crop performance: breeding and management practice. Improving wheat yield and WUE through breeding has been a major focus in semi-arid environments (Richards et al., 2002, 2010; Reynolds et al., 2005).

Management practices are equally important as breeding under water-limited conditions (Passioura and Angus, 2010; Richards et al., 2010). Improved crop management is responsible for a large portion of increased productivity under water-limited conditions (Anderson, 2010). Based on the long-term field studies in the SGP and NCP, we provided four case studies that are important for wheat water management in the two regions.

7.1 Soil and water conservation

Under semi-arid environmental conditions in the SGP, dryland wheat production is dependent on both soil water storage and growing season precipitation since precipitation only can meet part of seasonal ET requirement. Soil water storage at planting has long been a major focus in dryland wheat production (Musick et al., 1994; Stone and Schlegel, 2006). Musick et al. (1994) analysed the relationship between wheat yield and available soil water (ASW) at planting (1.8 m profile) based on 34 years of dryland wheat data in Bushland, Texas. There was a linear relationship between yield and ASW at planting ($Y = 0.0157X - 0.94$, $R^2 = 0.34$, $P < 0.001$). The linear regression resulted in a yield response of 1.57 kg m^{-3} to soil water storage in 1.8 m profile. In another word, increasing 1 mm of ASW at planting would lead to 15.7 kg ha^{-1} yield increase (Musick et al., 1994). Stone and Schlegel (2006) showed another linear relationship between ASW at emergence using 30-year dryland wheat data in Tribune, Kansas ($Y = 0.0098X + 0.828$, $R^2 = 0.32$, $P < 0.0001$), and resulted in a yield response of 9.8 kg ha^{-1} yield increase with additional 1 mm of ASW at emergence in 1.8 m profile.

The linear relationships between yield and ASW at planting or emergence emphasize the importance of preseason soil water storage and conservation of precipitation for dryland wheat production. For the linear regressions from Musick et al. (1994), the slope of the regression between yield and ASW at planting was much higher than that of the regression between yield and seasonal ET (1.57 vs. 1.22 kg m^{-3}). This suggests that ASW at planting is important for wheat plants to effectively use precipitation during growing season. The lower R^2 of regression between yield and ASW at planting or emergence (<0.40) indicated that wheat yield is affected by other factors. Seasonal precipitation is also significantly related to dryland wheat yield. Musick et al. (1994) showed that precipitation from October to June accounted for 55% dryland wheat yield variation in Bushland, Texas. Stone and Schelgel (2006) demonstrated that wheat yield increased 8.3 kg ha^{-1} with additional 1 mm of seasonal precipitation. Nevertheless, high yield in dryland wheat only can be achieved with both high ASW at planting and seasonal precipitation. In our recent dryland wheat studies, wheat in 2010–11 season was largely related to soil water storage and yield averaged only 0.65 Mg ha^{-1} because there was only 38 mm of seasonal precipitation (versus 250 mm in a regular season) (Xue et al., 2014). The dryland wheat in 2011–12 was mainly dependent on seasonal precipitation because of low soil water storage due to historic drought in 2011. As a result, wheat yield averaged 0.89 Mg ha^{-1} (Ajayi et al., 2016). In the 2009–10 season, dryland yield average 2.5 Mg ha^{-1} because of more ASW and seasonal precipitation as compared to other two seasons (Xue et al., 2014).

Soil water storage is determined by the amount of precipitation as well as the efficiency of precipitation storage during the fallow period. In the SGP, management practices for increasing soil water storage have focused on crop rotation, tillage systems and residue management (Musick et al., 1994; Jones and Popham, 1997; Stone and Schlegel, 2006; Stewart et al., 2010). Dryland wheat may be grown under different rotation systems such as continuous wheat (CW), wheat-fallow (WF) and wheat-sorghum-fallow (WSF). The tillage systems include conventional tillage and no-till with wheat residue mulch. Jones and Popham (1997) investigated the effects of rotation and tillage systems on dryland wheat and sorghum production and WUE in a 10-year study (1984–93) at Bushland, Texas. Among the different rotation systems, ASW at planting was always lower in CW system than in WF and WSF systems. The 10-year average ASW amounts at planting were 156, 212 and 205 mm for CW, WF and WSF, respectively. For the tillage systems, no-till significantly increased ASW at planting (199 mm) as compared to stubble mulch (183 mm). The precipitation storage efficiency (PSE) was generally low in Bushland, Texas, for wheat. Among the various treatments, WSF generally had higher PSE (about 17%) than WF (11%) and CW under stubble mulch (14%). However, no-till with wheat residue mulch in CW resulted in a higher PSE (20%) than WSF, indicating no-till is an important practice for improving PSE under the Great Plains environment (Jones and Popham, 1997). The above long-term study clearly demonstrated that WSF system under no-till provided significant benefits for improving wheat yield and WUE, as a result of increasing ASW at planting and PSE. On average, annual grain production was 1.8 Mg ha^{-1} for WSF (wheat and sorghum) but only 1.0 Mg ha^{-1} for CW (wheat only) (Jones and Popham, 1997). Norwood (1994) compared conventional and no-till tillage under WF and WSF systems. Although they did not find differences in wheat WUE between conventional tillage and no-till, less evaporation and runoff in no-till system promoted water moving to deeper soil profile in the WF and WSF systems. Stone and Schlegel (2006) sorted wheat yield and WUE data by tillage and showed that no-till significantly increased dryland wheat yield and WUE. The wheat yield response to ET was greater with no-till (1.38 kg m^{-3}) than with conventional tillage (0.86 kg m^{-3}). Stone and Schlegel (2006) also summarized the additional water increase in the soil profile during fallow period as a result of no-till, comparing with conventional tillage in the Great Plains. The soil water increase due to no-till ranged from 15 mm in SGP to 87 mm in central Great Plains, depending on rotation systems.

7.2 Irrigation and irrigation scheduling

Irrigation remains the most effective way to sustain high crop productivity. In the southern Great Plains, irrigation technology has changed significantly in last four decades, from furrow irrigation in early years (1950s–1970s) to current central pivot sprinkler and subsurface drip irrigation systems. The history and trends of irrigation research and development in the region have been reviewed in different eras from 1990s to recent time (Musick et al., 1990; Colaizzi et al., 2008; Evett et al., 2014). The details of irrigation history, economic impact, and research and development trends can be found in these three reviews. Based on these review papers, there are some general consensus for irrigation in the SGP. First, irrigation supplies from Ogallala Aquifer are declining continuously. Second, irrigation efficiency has been increased significantly from 1950s to current time. Third, the future challenge will be how to more efficiently use reduced amount of irrigation in centre pivot system. However, sustaining irrigation is crucial to sustainable crop production in the region because irrigation can significantly increase WUE as compared to dryland (Evett et al., 2014).

In order to increase the efficiency of the irrigation application, irrigation should be scheduled using measurements of crop ET and/or soil moisture depletion. Advanced irrigation scheduling to enhance the WUE includes management of both water and soils. Consequently, irrigation scheduling is largely related to ET when striving to increase the seasonal crop WUE. ET for a specific crop is normally calculated using a crop-specific coefficient (Kc) and the reference ET (ET_o). ET_o is the calculated ET from a reference crop (turf grass or alfalfa) using weather station data including temperature, relative humidity, precipitation, wind speed and solar radiation. Irrigation is also scheduled according to changes in soil moisture. In the past, using ET was not practiced for by many irrigators but ET today can be readily calculated from meteorological data directly from local weather stations. State and commercial service providers in many regions of winter production make available daily ET estimates as part of their outreach programs or consulting service. Recently, estimates of daily and forecasted weekly reference ET are now being made available by the U.S. National Weather Service. International plans for these same estimates are being considering for global locations and applications in the near future. Soil moisture sensors provide irrigators the knowledge of soil moisture fluxes in the root zone, which provides information of when to initiate and terminate irrigation and ability to create a seasonal soil water balance.

The irrigation system of choice must be able to efficiently deliver and apply the amount of water needed to meet the crop water requirement. Knowledge of an irrigation system's efficiency is critical to ensure the crop is uniformly irrigated at the targeted level. Efficiency describes fraction of water stored in the root zone compared to the volume of water delivered by the system, which accounts for drainage below the root zone, evaporation and the uniformity of the application. Irrigation systems used to irrigate wheat include pressurized systems (sprinkler and subsurface drip irrigation) and surface irrigation (furrow and flooded borders), but according to the 2013 Nation Agricultural Statistics Service 2013 survey, 97% of irrigation is applied with a pressurized system of which are predominately sprinkler. For subsurface drip irrigation systems, on expansive soils, very deep cracks develop as wheat matures and dries down. Most producers have discovered that rodents invade the cracks and damage the drip tape resulting in the need for significant repairs. Sprinkler systems include centre pivot, linear move, wheel roll and travelling gun sprinklers. System efficiencies vary between systems; subsurface drip systems generally operate at efficiencies greater than 90% and typically can be in the 95 to 98% range, but the application efficiencies of surface and sprinkler systems can vary greatly. Sprinkler application efficiencies vary generally from 70% to 95%, depending on sprinkler nozzle design, height of the applicator or nozzles, temperature and wind conditions. Surface systems are generally the lowest efficiency system. The drip and center pivot systems offer the greatest opportunity to implement advanced irrigation scheduling management in winter wheat production.

7.3 Deficit irrigation

In the U.S. SGP, with the declining irrigation water supplies from Ogallala Aquifer and increasing energy costs, application of less irrigation water than the plants require for high yield will be the primary practice in the future for irrigated wheat production. Deficit irrigation has been studied and practiced for over three decades and shown to be a viable management practice for improving yield and WUE (Eck, 1988; Musick, et al., 1994; Schneider and Howell, 2001; Xue, et al., 2003, 2006). Deficit irrigation is defined as the application of less water than it is required for full ET and maximum yield (English et al., 1990; Musick, et al. 1994). It is called by a variety of other names such as partial irrigation, regulated deficit irrigation, ET deficit irrigation and limited irrigation (English et al., 1990; Kang et al., 2000; Fereres and Soriano, 2007). The overall goal of deficit irrigation is to increase WUE, either by reducing irrigation frequency or by eliminating the least productive irrigations. Management of deficit irrigation is very different from full irrigation management. Rather than working to minimize crop water stress, plants must be allowed to certain levels of water stress (English et al., 1990; Zhang et al., 1998).

Compared to dryland wheat, deficit irrigation significantly increased grain yield and WUE (Xue et al., 2012). Table 1 showed the responses of wheat yield, WUE, ET, irrigation water-use efficiency (IWUE), and harvest index (HI) to different irrigation levels (from deficit to full irrigation) in two growing seasons under a centre pivot irrigation system. Irrigation application significantly increased wheat yield and WUE as compared to dryland treatment. For example in 1997–98 season, the dryland yield was only 2.5 Mg ha^{-1} and WUE only 0.57 kg m^{-3}. However, yields in irrigated treatments ranged from 4.5 to 7.2 Mg ha^{-1} and WUE from 0.85 to 1.06 kg m^{-3}. Among the irrigation treatments, deficit irrigation (50% ET requirement) resulted in highest WUE and IWUE. For 50% ET treatment, yield was 86–95% of full irrigation (100% ET) (Schneider and Howell, 2001). In another field study, Xue et al. (2006) showed that deficit irrigation of 100 mm at booting stage increased yield by 46% and WUE by 23% as compared to dryland treatment. Compared to full irrigation of 400 mm, a deficit irrigation of 220 mm at jointing and anthesis achieved 84% of the yield at full irrigation and resulted in 45% irrigation water savings. Recently, we used long-term weather data and a simulation model further confirmed these results (Attia et al., 2016). In the NCP, Zhang et al. (1998) showed that wheat yield at one-irrigation (80 mm) was 84–100% of the yield at four-irrigation (320 mm). As a result, WUE was increased to 1.55 kg m^{-3} at one-irrigation treatment from 1.22 kg m^{-3} at four-irrigation treatment

Table 1 Irrigation amounts, yield, evapotranspiration (ET), water-use efficiency (WUE), irrigation water-use efficiency (IWUE) and harvest index (HI) under different irrigation levels in two growing seasons under sprinkler irrigation system (Schneider and Howell, 2001, used with permission)

Crop Season	Irrigation % full ET	mm	Yield Mg ha^{-1}	ET mm	WUE kg m^{-3}	IWUE kg m^{-3}	HI
1997–98	0%	0	2.46d	432	0.57c	–	0.16d
	25%	109	4.48c	527	0.85b	1.93a	0.25c
	50%	213	6.79b	638	1.06a	2.13a	0.29a
	75%	319	6.43b	745	0.86b	1.30b	0.27b
	100%	419	7.17a	824	0.87b	1.19b	0.29a
1998–99	0%	0	4.08e	557	0.73b	–	0.28c
	25%	73	4.93d	622	0.80a	1.36a	0.29b
	50%	144	5.85c	697	0.84a	1.26a	0.30ab
	75%	221	6.30b	782	0.81a	1.01b	0.31a
	100%	297	6.79a	825	0.82a	0.88b	0.32a

Growing season precipitation: 212 mm in 1997–98; 334 mm in 1998–99.

(Zhang et al., 1998). More recent studies confirmed the above results in the NCP (Qiu et al., 2008; Zhang et al., 2011).

Many field studies have been conducted in the SGP and NCP to elucidate the mechanisms for increased yield and WUE under deficit irrigation conditions. First, maintaining high ASW at planting and allowing soil drying at an early stage are important for successful practice of deficit irrigation. Xue et al. (2003) showed that mild soil water stress in the spring promoted root growth and development. For example, root length density along the soil profile was higher in dryland than in irrigated treatment at booting. With the deep root system, irrigation can be delayed until booting stage or anthesis if only one-irrigation is allowed (Xue et al., 2003). Similarly, Zhang et al. (1998) and Zhang et al. (2011) also reported that mild water stress at early stage led to a relatively deeper root system in NCP environment.

Second, irrigation must be applied at critical stages for wheat yield determinations when only limited irrigation is allowed. Critical growth stages for irrigating winter wheat generally occur from early spring growth to early grain development in the SHP. When only one-irrigation allowed, irrigation between jointing and anthesis resulted in same yield and WUE. However, irrigation at grain filling did not increase yield and WUE (Xue et al., 2003, 2006). Irrigation at critical stage also reduced the severity of soil water stress and allowed plants to use more water during grain filling (Zhang et al., 1998; Xue et al., 2003, 2006). Xue et al. (2003, 2006) demonstrated that one-irrigation at booting stage maintained high soil water content and significantly reduced water stress at anthesis. As a result, plants in one-irrigation at booting stage had higher photosynthetic rate and stomatal conductance than dryland and one-irrigation at jointing (Xue et al., 2006). Similar results were also found in the NCP (Zhang et al., 1998, 2011).

Third, increased yield and WUE under deficit irrigation is related to increased HI. HI is determined during grain filling by both current photosynthesis and remobilization of pre-anthesis carbon reserve from stems. For the maintenance of current photosynthesis to meet the carbohydrate supply, higher photosynthesis rate and longer green leaf area duration are advantageous under drought conditions. Zhang et al. (2011) showed that deficit irrigation changed wheat canopy structure and increased the proportion of non-leaf organs for photosynthesis. As a result, canopy photosynthetic capacity was improved during grain filling and grain yield was increased under deficit irrigation. When the photosynthesis during grain filling is reduced by drought stress, remobilization of carbon reserves can be important to grain filling. The contribution of remobilized carbon reserves to grain yield in wheat varied from 5% to 90%, depending on the environmental conditions (Asseng and van Herwaarden, 2003; Xue et al., 2006). The increased HI under appropriate deficit irrigation was due to increase in both current photosynthesis and the remobilization of pre-anthesis carbon reserves (Xue et al., 2006).

7.4 Genetic improvement of drought tolerance and WUE

Since drought stress is inevitable during wheat growing season under water-limited conditions, particularly in the U.S. SGP, improving wheat drought tolerance through breeding continuously remains an important part of overall crop improvement (Xue et al., 2012, 2014). Wheat yield has been increasing through breeding in the SGP. For example, regional cultivar trials in the Texas High Plains showed that newly released cultivars such as TAM 111 (released in 2003) and TAM 112 (released in 2005) consistently had higher yields than a historic check, TAM W-101 (released in 1969) under both dryland and irrigated conditions. On average, the yield benefits with newly released cultivars ranged from 15% to 35% over check cultivars (Bean, 2010, 2011; Xue et al., 2014). The yield benefits from two newly released cultivars, TAM 111 and TAM 112, were particularly beneficial in the SGP. TAM 111 and TAM 112 were top two cultivars planted in the Texas High Plains and western Kansas acreage in 2011 and 2012 (NASS, 2011, 2012a, b). Despite the genetic improvement in yield, yield is still the primary trait used in cultivar selection for drought tolerance. Therefore, better understanding physiological mechanisms of drought tolerance and identification of plant traits are important for developing cultivars with improved drought tolerance and WUE.

In last few years, we have been conducting both field and greenhouse experiments to understand the physiological basis of yield determination, drought tolerance and WUE in wheat genotypes. We have mainly focused on cultivars and genotypes developed by Texas A&M AgriLife Research. In a 2-year field study conducted in Bushland, Texas, Xue et al. (2014) investigated the physiological basis of yield determination and WUE in 10 genotypes under dryland and irrigated conditions. The results showed that the newer cultivars such as TAM 111 and TAM 112 had higher yield, biomass, WUE and WUE_{bm} than a relatively older cultivar (TAM 105) under drought. Moreover, TAM 111 and TAM 112 were more drought tolerant than TAM 105 and another drought susceptible line (TX86A5606). The WUE or WUE_{bm} was determined by yield or biomass as genotypic differences in ET were not significant. Biomass at anthesis significantly contributed to increased yield under drought. For dryland wheat, remobilization of stem carbon reserves also contributed to yield. Foulkes et al. (2002) showed that winter wheat cultivars with more drought tolerance had higher HI and more carbon remobilization under drought in the United Kingdom. Our further studies demonstrated that higher yield in drought tolerant cultivars under dryland was related to cooler canopy temperature and more effective use of soil water in deeper profile. In addition, the two drought tolerant cultivars (TAM 111 and TAM 112) had different mechanisms to respond drought. TAM 111 performed well with a relatively high soil moisture conditions at early stage. However, TAM 112 can sustain longer and severe drought conditions (Pradhan et al., 2014a,b). In the NCP of China, Zhang et al. (2010) showed that newer cultivars not only had higher yield but also required less irrigation water, indicating that newer cultivars are more drought tolerant than older cultivars.

Reddy et al. (2014) further conducted greenhouse study to understand the differences in the physiological and transcriptomic responses of TAM 111 and TAM 112 to drought stress during grain filling. Whole-plant data indicated that TAM 112 used more water, produced more biomass and grain yield under water stress compared to TAM 111. Leaf-level data at the grain filling stage indicated that TAM 112 had elevated abscisic acid (ABA) content and reduced stomatal conductance and photosynthesis as compared to TAM 111. Sustained drought stress during the grain filling also resulted in greater flag leaf transcriptome changes in TAM 112 than TAM 111. The two cultivars had very different transcripts associated with photosynthesis, carbohydrate metabolism, phytohormone metabolism and other dehydration responses under drought stress.

Based on these studies, some important physiological traits (e.g. biomass at anthesis, canopy temperature, carbon reserve remobilization and ABA) associated with drought tolerance and WUE can be identified under water-limited conditions in the SGP. Further research is needed to evaluate these traits in large breeding populations and identify QTLs and molecular markers. Because the effective use of soil water is important for ET and yield in the SGP, a better understanding root traits will also be needed in the near future.

8 Future trends and conclusion

Although progress has been made to improve wheat yield and WUE in last few decades in the U.S. SGP, the current WUE level is still low (<1.2 kg m^{-3}) as compared to other winter wheat production areas such as NCP (about 2.0 kg m^{-3}). Therefore, developing better management strategies to improve yield and WUE is still a challenge for agricultural scientists, particularly under changing climate scenario. For dryland wheat production in the SGP, management practices (e.g. no-till, residue mulching and rotation) have resulted in significant gain in soil water storage during fallow periods over the last four to five decades. Future research must address how to efficiently use soil water. Since drought stress is inevitable during wheat growing season, improving drought tolerance through breeding will be an increasingly important part of overall crop improvement. A better understanding of crop response to drought stress and identification of plant traits will lead to development of improved germplasm and cultivars in the region. Over the years, there have been some progresses for better understanding drought tolerance and WUE in winter wheat under drought in the SGP. However, further studies are still needed to identify more traits related to improved drought tolerance and WUE. Currently, the major challenge is high-throughput field phenotyping as more and more genotypes are readily available by genotyping and genomic technologies.

Declining irrigation water continues to challenge irrigated wheat production in the SGP. Although several management factors (improved cultivars, fertilization and pests control, etc.) contributed to wheat yield and WUE improvement, irrigation has played a vital role to increase wheat yield and WUE in the region. Irrigated wheat yields can be 2–4 times higher than dryland yields. Therefore, irrigation will be an important management practice for a long time. As irrigation water becomes more limiting, deficit irrigation will be the primary practice in the future for irrigated wheat production. This is also true in NCP. Currently, irrigation frequency has reduced to 1–2 irrigations per season as compared to historically 4 irrigations during winter wheat growing season. Emerging precision application tools and irrigation systems control technologies currently being developed will enhance and compliment the understanding of practices and genetics as well yield and WUE in winter wheat production.

9 Where to look for further information

More information regarding water management and drought tolerance in wheat can be found by searching the key words in scholar.google.com.

The general crop water management topics also can be found in the websites of the departments of agronomy or soil & crop sciences in many universities such as Texas A&M University, Kansas State University, University of Nebraska and University of California, to name a few.

Popular journals: *Crop Science, Agronomy Journal, Field Crops Research, Functional Plant Biology, European Journal of Agronomy, Agricultural and Water Management, Irrigation Science, Journal of Experimental Botany, Advances in Agronomy* and so on.

Websites: plantstress.com.

Conferences: American Society of Agronomy, Crop Science Society of America, Soil Science Society of America International Annual meeting; American Society of Plant Biologists Annual Meeting; Inter-Drought Conference; American Society of Agricultural and Biosystems Engineering Annual Meeting.

10 References

AgriPartners (2007). Irrigation and Cropping Demonstrations. http://amarillo.tamu.edu/amarillo-center-programs/agripartners/.

Ajayi, S., Reddy, S. K., Gowda, P. H., Xue, Q., Rudd, J. C., Pradhan, G., Liu, S., Stewart, B. A. Biradar, C. and Jessup, K. E. (2016). Spectral reflectance models for characterizing winter wheat genotypes. *Journal of Crop Improvement* 30: 176–95. DOI:10.1080/15427528.2016.1138421.

Anderson, W. K. (2010). Closing the gap between actual and potential yield of rainfed wheat. The impacts of environment, management and cultivar. *Field Crops Research* 116: 14–22.

Attia, A., Rajan, N., Xue, Q., Ibrahim, A. and Hays, D. (2015). Application of DSSAT-CERES-Wheat model to simulate winter wheat response to irrigation management in the Texas High. *Agricultural Water management* 165: 50–60.

Asseng, S. and van Herwaarden, A. F. (2003). Analysis of the benefits to yield from assimilates stored prior to grain filling in a range of environments. *Plant Soil* 256: 217–29.

Bean, B. W. (2010). 2010 wheat variety trials conducted in the Texas and New Mexico High Plains. http://amarillo.tamu.edu/files/2010/11/Wheat-Variety-Descriptions-2010-7_26.pdf.

Bean, B. W. (2011). 2011 wheat variety trials conducted in the Texas and New Mexico High Plains. http://amarillo.tamu.edu/files/2010/11/Wheat-Variety-Trial-Summary-Final2011.pdf.

Blum, A. (2009). Effective use of water (EUW) and not water-use efficiency (WUE) is the target of crop yield improvement under drought stress. *Field Crop Research* 112:119–23.

Bodner, G., Nakhforoosh, A. and Kaul, H. (2015). Management of crop water under drought: a review. *Agronomy for Sustainable Development* 35: 401–42.

Bringle, R. W. and Curtis, B. C. (1987). Wheat worldwide. In Hey, E. G. (ed.), *Wheat and Wheat Improvement*, Second Edition, pp. 1–32. ASA-CSSA-SSSA, Madison, WI.

Colaizzi, P. D., Gowda, P. H., Marek, T. H. and Porter, D. O. (2008). Irrigation in the Texas High Plains: A brief history and potential reductions in demand. *Journal of Irrigation and Drainage Engineering* 58. doi:10.1002/ird.418.

Eck, H. V. (1988). Winter wheat response to nitrogen and irrigation. *Agronomy Journal* 80: 902–8.

English, M. J, Musick, J. T. and Murty, V. V. N. (1990). Deficit irrigation. In *Management of Farm Irrigation Systems*. American Society of Agricultural Engineers, St. Joseph, MI, pp. 631–63.

English, M. J. (1990). Deficit irrigation: Observations in the Columbia Basin. *Journal of Irrigation and Drainage Engineering* 116: 413–26.

Evett, S. R., Colaizzi, P. D., Susan, A. O'Shaughnessy, S. A., Lamm, F. R., Trout, T. J. and Kranz, W. L. (2014). The future of irrigation on the U.S. Great Plains. *Proceedings of the 26th Annual Central Plains Irrigation Conference*, Burlington, CO., 25–26 February 2014. Available from CPIA, 760, N. Thompson, Colby, Kansas, USA.

Fanin, B. (2011). Texas agricultural drought losses reach record $5.2 billion. Texas AgriLife News, 17 August 2011. http://agrilife.org/today/2011/08/17/texas-agricultural-drought-losses-reach-record-5-2-billion/.

Fereres, E., and Soriano, M. A. (2007). Deficit irrigation for reducing agricultural water use. *Journal of Experimental Botany* 58 (2): 147–59.

Fischer, R. A., Byerlee, D. and Edmeades, G. O. (2014). Crop yields and global food security: will yield increase continue to feed the world? *Canberra: Australian Centre for International Agricultural Research*. http://aciar.gov.au/publication/mn158

Foulkes, M. J., Scott, R. K. and Sylvester-Bradley, R. (2002). The ability of wheat cultivars to withstand drought in UK conditions: formation of grain yield. *Journal of Agricultural Science* 138: 153–69.

Graybosch, R. A. and Peterson, C. J. (2010). Genetic improvement in winter wheat yields in the Great Plains of North America, 1959–2008. *Crop Science* 50: 1882–90.

Hao, B., Xue, Q., Zhang, Y. H., Stewart, B. A. and Wang, Z. M. (2014). Deficit irrigation in winter wheat-U. S. Southern High Plains and North China Plain. *Journal of Arid Land Studies* 24–1: 129–32.

Howell, T. A. (2001). Enhancing water use efficiency in irrigated agriculture. *Agronomy Journal* 93: 281–9.

Howell, T. A., Tolk, J. A., Evett, S. R., Copeland, K. S. and Dusek, D. A. (2007). Evapotranspiration of deficit irrigated sorghum and winter wheat. *Proceedings USCID 4th International Conference*, pp. 223–39.

Howell, T. A., Steiner, J. L., Schneider, A. D. and Evett, S. R. (1995). Evapotranspiration of irrigated winter wheat: Southern High Plains. *Transactions of the ASAE* 38: 745–59.

Howell, T. A. (1990). Grain, dry matter yield relationships for winter wheat and grain sorghum – Southern High Plains. *Agronomy Journal* 82: 914–18.

Howell, T. A., Steiner, J. L., Schneider, A. D., Evett, S. R. and Tolk, J. A. (1997). Seasonal and maximum daily evapotranspiration of irrigated winter wheat, sorghum, and corn Southern High Plains. *Transactions of the ASAE* 40: 623–34.

Jones, O. R. and Popham, T. W. (1997). Cropping and tillage systems for dryland grain production in the Southern High Plains. *Agronomy Journal* 89: 222–32.

Kang, S., Zhang, L., Liang, Y., Hu, X., Cai, H. and Gu, B. (2002). Effects of limited irrigation on yield and water use efficiency of winter wheat in the Loess Plateau of China. *Agricultural Water Management* 55: 203–16.

Liu, C., Zhang, X. and Zhang, Y. (2002). Determination of daily evaporation and evapotranspiration of winter wheat and maize by large-scale weighing lysimeter and micro-lysimeter. *Agricultural and Forest Meteorology* 111: 109–20.

Mo, X., Liu, S., Lin, Z. and Guo, R. (2009). Regional crop yield, water consumption and water use efficiency and their responses to climate change in the North China Plain. *Agriculture, Ecosystems and Environment* 134: 67–78.

Musick, J. T and Porter, K. B. (1990). Wheat. In Stewart, B. A. and Nielson, D. R. (eds), *Irrigation of Agricultural Crops*, Agron. Monogr. 30. Madison: ASA-CSSA-SSSA, pp. 597–638.

Musick, J. T., and Dusek, D. A. (1980). Planting date and water deficit effects on development and yield of irrigated winter wheat. *Agronomy Journal* 72: 45–52.

Musick, J. T., Pringle, F. B., Harman, W. L. and Stewart, B. A. (1990). Long-term irrigation trends–Texas High Plains. *Applied Engineering In Agriculture* 6: 717–24.

Musick, J. T., Jones, O. R., Stewart, B. A. and Dusek, D. A. (1994). Water–yield relationships for irrigated and dryland wheat in the U.S. Southern Plains. *Agronomy Journal* 86: 980–6.

NASS. (2011). 2010 Texas Wheat Variety Survey Results. http://www.nass.usda.gov/Statistics_by_State/Texas/Publications/Crop_Reports/Wheat/twheat_var.htm.

NASS. (2016). Winter Wheat: Production Acreage by County https://www.nass.usda.gov/Charts_and_Maps/Crops_County/ww-pr.

NASS. (2012a). Texas 2012 Wheat Variety Results. http://www.nass.usda.gov/Statistics_by_State/Texas/Publications/tx_wheat_varieties.pdf.

NASS. (2012b). Wheat Varieties. http://www.nass.usda.gov/Statistics_by_State/Kansas/Publications/Crops/Whtvar/whtvar12.pdf.

Norwood, C. (1994). Profile water distribution and grain yield as affected by cropping system and tillage. *Agronomy Journal* 86: 558–63.

Passioura, J. B. and Angus, J. F. (2010). Improving productivity of crops in water-limited environments. *Advances in Agronomy* 106: 37–75.

Passioura, J. B. (1977). Grain yield, harvest index and water use of wheat. *Journal of the Australian Institute of Agricultural Science* 43: 117–20.

Pradhan, G., Xue, Q., Liu, S., Rudd, J. C. and Jessup, K. E. (2014a). Effective use of soil water contributed to high yield in wheat in the U.S. Southern High Plains. *Journal of Arid Land Studies* 24–1: 153–6.

Pradhan, G., Xue, Q., Liu, S., Rudd, J. C., Jessup, K. E. and Mahan, J. R. (2014b). Cooler canopy temperature contributed to higher yield in new drought tolerant cultivars. *Crop Science* 54: 2275–84.

Qiu, G., Wang, L., He, X., Zhang, X., Chen, S., Chen, J. and Yang, Y. (2008). Water use efficiency and evapotranspiration of winter wheat and its response to irrigation regime in the north China plain. *Agricultural and Forest Meteorology* 148: 1848–59.

Reddy, S. K., Liu, S., Rudd, J. C., Xue, Q., Payton, P., Finlayson, S. A., Mahan, J., Akhunova, A., Holalu, S. V. and Lu, N. (2014). Physiology and transcriptomics of water-deficit stress responses in wheat cultivars TAM 111 and TAM 112. *Journal of Plant Physiology* 171: 1289–98.

Reynolds, M. P., Mujeeb-Kazi, A. and Sawkins, M. (2005). Prospects for utilising plant-adaptive mechanisms to improve wheat and other crops in drought- and salinity-prone environments. *Annals of Applied Biology* 146: 239–59.

Richards, R. A., Rebetzke, G. J., Condon, A. G. and van Herwaarden, A. F. (2002). Breeding opportunities for increasing the efficiency of water use and crop yield in temperate cereals. *Crop Science* 42: 111–21.

Richards, R. A., Rebetzke, G. J., Watt, M., Condon, A. G., Spielmeyer, W. and Dolferus, R. (2010). Breeding for improved water productivity in temperate cereals: phenotyping, quantitative traits loci, markers and the selection environment. *Functional Plant Biology* 37: 85–97.

Schneider, A. D. and Howell, T. A. (2001). Scheduling deficit irrigation with data from an evapotranspiration network. *Transaction of the ASAE* 44: 1617–23.

Sinclair, T. R. and Muchow, R. C. (2001). System analysis of plant traits to increase grain yield on limited water supplies. *Agronomy Journal* 93: 263–70.

Stewart, B. A. and Peterson, G. A. (2015). Managing green water in dryland agriculture. *Agronomy Journal* 107: 1544–53.

Stewart, B. A., Baumhardt, R. L. and Evett, S. R. (2010). Major advances of soil and water conservation in the U.S. Southern Great Plains. Zoebeck, T. M., Schillinger, W. F.,editors. Soil and Water Conservation Advances in the United States. Special Publication 60. Madison, WI: Soil Science Society of America, Inc., pp. 103–29.

Stone, L. R. and Schlegel, A. J. (2006). Yield–water supply relationships of grain sorghum and winter wheat. *Agronomy Journal* 98: 1359–66.

USDA-ERS. (2015). U.S. share of world wheat exports continues to decline. http://www.ers.usda.gov/data-products/chart-gallery/detail.aspx?chartId=52238.

Wang, B., Zhang, Y., Hao, B., Xu, X., Zhao, Z., Wang, Z. and Xue, Q. (2016). Grain yield and water use efficiency in extremely-late sown winter wheat cultivars under two irrigation regimes in the north china plain. *PLoS ONE* 11(4): e0153695. doi:10.1371/journal.pone.0153695.

Wang, F. Z.e, Sayre, K., Li, S., Si, J., Feng, B. and Kong, L. (2009). Wheat cropping systems and technologies in China. *Field Crops Research* 111: 181–8.

Xue, Q., Liu, W. and Stewart, B. A. (2012). Improving wheat yield and water use efficiency under semi-arid environment – The US Southern Great Plains and China's Loess Plateau. In R. Lal and B. A. Stewart (eds), *Advances in Soil Science: Soil Water and Agronomic Productivity*. Taylor & Francis.

Xue, Q., Zhu, Z., Musick, J. T., Stewart, B. A. and Dusek, D. A. (2003). Root growth and water uptake in winter wheat under deficit irrigation. *Plant Soil* 257: 151–61.

Xue, Q., Zhu, Z., Musick, J. T., Stewart, B. A. and Dusek, D. A. (2006). Physiological mechanisms contributing to the increased water-use efficiency in winter wheat under deficit irrigation. *Journal of Plant Physiology* 163: 154–64.

Zhang, H., Wang, X., You, M. and Liu, C. (1999). Water – yield relations and water-use efficiency of winter wheat in the North China Plain. *Irrigation Science* 19: 37–45.

Zhang, X., Pei, D. and Hu, C. (2003). Conserving groundwater for irrigation in the North China Plain. *Irrigation Science* 21: 159–66.

Zhang, Y., Zhang, Y., Wang, Z. and Wang, Z. J. (2011). Characteristics of canopy structure and contributions of non-leaf organs to yield in winter wheat under different irrigated conditions. *Field Crops Research* 123: 187–95.

Zhang, J. H., Sui, X. Z., Li, B., Su, B. L., Li, J. M. and Zhou, D. X. (1998). An improved water-use efficiency for winter wheat grown under reduced irrigation. *Field Crops Research* 59: 91–8.

Zhang, X., Chen, S., Sun, H., Wang, Y. and Shao, L. (2010). Water use efficiency and associated traits in winter wheat cultivars in the North China Plain. *Agriculture Water Management* 97: 1117–25.

Chapter 4

Advances in irrigation techniques for rice cultivation

D. S. Gaydon, CSIRO Agriculture, Australia

1 Introduction
2 Water-saving measures
3 Scale-dependency of water productivity and water savings
4 Aerobic rice as a water-saving measure
5 Alternate wetting and drying (AWD) as a water-saving measure
6 Saturated soil culture (SSC) as a water-saving measure
7 Case study: water-saving irrigation in southeast Australia
8 Future trends and conclusion
9 Where to look for further information
10 References

1 Introduction

Increasing scarcity of fresh water resources is a growing concern in many major food-producing areas of the world, including some of the major rice-growing areas in Asia and Africa (Godfray et al., 2010; Rijsberman, 2006). Projected future increases in temperatures are also likely to exacerbate the issue of crop water stress (IPCC, 2014). This, combined with the imperative to produce more food for a growing global population, is driving a focus globally on increasing water productivity in irrigated agriculture, or, put more simply, growing more food with less water (Keating et al., 2010).

Rice is one of the key elements of global food security, being the staple food for most of the world's population, particularly people in low and lower-middle-income countries. Nearly 90% of the global rice crop is produced by Asian farmers (Devendra and Thomas, 2002), and some of the facts associated with production of this food source are truly staggering. For example, rice farming is the single largest use of land for producing food on Earth (with some 696 million tons produced in 2010 from around 158 million hectares), yet only 7% of all rice is exported from its country of origin. Rice is produced along Myanmar's Arakan Coast, where the growing season average rainfall is greater than 5100 mm; in places like Australia's Riverina and parts of Africa where the growing season rainfall may be less than 100 mm; at the Upper Sindh in Pakistan with average season temperatures of 33°C; and in parts of Japan where the mean growing season temperature falls as low as 17°C (GRISP, 2013).

With such huge variation in rice-growing environments and landscapes, not surprisingly there are a wide range of cultivation and irrigation practices. The majority of the world's rice is grown under irrigated conditions in which the fields are flooded from planting to harvest. Rice land receives 35–45% of all the world's irrigation water (which itself uses some 70% of all the world's developed water resources) (Bouman, 2013; Barker et al., 1998). In view of the importance of rice to humanity and the current and growing concerns about future water scarcity, achieving water savings in irrigated rice production has become one of the key research challenges for the rice-growing world.

There are a number of potential research avenues being pursued to achieve such water savings without sacrificing rice yields. Essentially the aim is to increase irrigated water productivity – the amount of rice produced per unit of irrigation water applied. These avenues include (i) breeding of short-duration, high-yielding modern rice varieties which require less water due to less time in the field (Khush, 1995); (ii) breeding of varieties which yield better in harsh environments (salinity, submergence, heat, cold, drought) (Zhang, 2007); (iii) breeding of varieties better suited to new water-saving irrigation practices (for example deep-rooted genes for more water-stressed conditions) (iv) precision agriculture, site-specific N management and improved crop establishment methods to increase yields and minimize wasted resources (Y. Yao et al., 2012; Peng et al., 2010; Kumar and Ladha, 2011); (v) land-forming and Conservation Agriculture practices which seek to improve soil health and system sustainability (Hobbs et al., 2008); and (vi) improved irrigation techniques which minimize unproductive water losses in the cropping system. In the future, successful water-saving practices will require continual adaptation to climatic changes and other emerging challenges, such as growing pressure to utilise marginal lands (Borlaug, 2007).

Other chapters in this publication are addressing the first four of the above avenues, and many of these focus on increasing water productivity by increasing yields as the primary focus. Although in reality many of these solutions work together in a system context (Tuong et al., 2005), this chapter will restrict itself to consideration of the key techniques for improving rice water productivity through enhanced irrigation practices and aiming to reduce irrigation water use in rice cultivation.

2 Water-saving measures

Before moving into a literature review on irrigation methods for water savings in rice production, a consideration of the definition of 'water productivity' (WP) is warranted to avoid confusion. A general definition of WP is:

WP = rice produced (amount or value)/**water used** (mm)

From the perspective of a crop physiologist or a breeder, the 'water used' component would refer to the amount of water transpired by the crop, leading to the 'transpired water productivity' (WPT – a purely physiological term). However from the perspective of the farmer or a field-scale irrigation planner, the 'water used' is a broader term, referring to the total irrigation water applied to the crop. They are interested in extracting as much value as possible out of every drop of available irrigation water, and maximising the 'irrigated water productivity' (WPI – which encompasses water used in crop transpiration, evaporation from the soil or pond surface, and also any losses of water from the field in the form of seepage under/through bunds, deep drainage and runoff). Further, at a regional scale the water resources planner will be more interested in the productivity of all regional water resources (WPTWI) including diverted or pumped irrigation water, rainfall, plus any deep drainage or runoff waters which are re-couped somewhere else in the system and re-used (Humphreys et al., 2010; Tuong et al., 2005).

As this review is on advances in irrigation techniques for rice cultivation, the focus will primarily be on methods and practices for saving irrigation water at the field scale (WPI); however considerations at the larger regional scale (WPTWI) will be reflected upon when relevant.

3 Scale-dependency of water productivity and water savings

The scale-dependency of water productivity and water savings has long been recognised (Molden et al., 2003). Water applied at a *field scale* is stored in the soil matrix, transpired to the atmosphere via the crop, lost by evaporation directly from the soil or water surface, or lost from the system via a number of other possible mechanisms (deep drainage below the reach of crop roots, lateral flows out of the field which include surface runoff or seepage under the bunds, etc.). Water applied in excess of crop needs results in an increasing proportion of these losses (Fig. 1; Fereres and Soriano, 2007). Technologies such as Conservation Agriculture (Hobbs, 2007) may lead to decreased water losses due to an increased capacity of the soil to store water (improved structure), and reductions in evaporation via use of surface mulches.

There is a general positive relationship between crop yield and water transpired by the crop (Perry et al., 2009); hence farmers may seek to minimise such losses from their fields in order to maximise their WPI. However, at a *larger scale*, the 'losses' of an individual farmer may be recouped by other users in the system (or even the farmer himself) and productively used, therefore not really constituting true losses. An example of this phenomenon is applied irrigation water lost as deep drainage from a field, entering the regional water table, to be once again pumped to the surface and used for irrigation and productive crop transpiration either at the same location or further afield (Humphreys et al., 2010; Paydar et al., 2009).

In Punjab, India, for example, declining water table depth has been a major concern for decades, with over-exploitation of water resources (particularly groundwater pumping) often being blamed (Jalota et al., 2007). Numerous water-saving practices have been mooted to save irrigation water and contribute to arresting the decline (such as laser land levelling, replacing rice with other crops, alternate wetting and drying (AWD) water management in rice, zero-till wheat, raised beds, delayed rice transplanting, shorter duration rice varieties, etc.). These irrigation water-saving practices may achieve water savings at the *field scale*, however those savings a likely due to reductions in deep drainage in many instances, with little effect on evapotranspiration (ET). Because deep drainage is returned to the water-table and not lost from the system in this region, reducing deep drainage alone is hence unlikely to have the required impact on water-table decline. It is important to note that reducing deep drainage may have major additional benefits in Punjab and is not irrelevant – for example reduced energy consumption for groundwater pumping and reduced pollution of the groundwater via nutrient leaching – however, this alone will not address the key environmental crisis of water-table decline. The irrigation water-saving options which primarily reduce ET are the ones that are most likely to succeed in addressing the problem in Punjab (Humphreys et al., 2010) as evaporative losses are the *true* losses in this system at the regional scale, and are driving the water-table decline. In other hydrogeological systems, the picture may be quite different – deep drainage losses may be genuine losses, un-recoupable by irrigators, and hence any savings in both ET and deep drainage may constitute true irrigation water savings in those systems.

Another aspect of scale which needs addressing in a review like this is the concept of water-limited versus land-limited production environments (Gaydon et al., 2012a). Water-savings are not a universally desirable aim in production systems – water savings do not in themselves allow the growth of more rice on *a given field* (Bouman et al., 2001). Rice is very sensitive to water stress, and attempts to reduce water inputs may result in yield reduction (Tuong et al., 2005). If the water savings from one field cannot be diverted to produce additional rice at another field (in other words, expanding the area of production), then an overall reduction in rice production may result. Field-level WP and yield can only be increased concurrently by water-saving irrigation plus raising crop yields through varietal improvement or better agronomy (fertiliser management, weed and pest control, etc.). Total rice production increases from water savings can occur by using the water saved

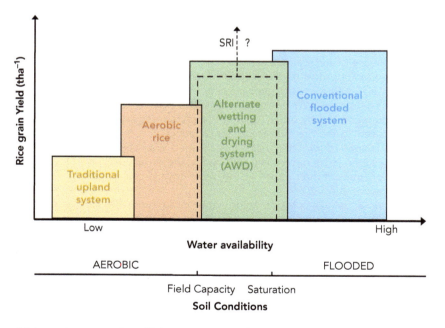

Figure 1 Schematic presentation of irrigation system and rice yields as a function of water availability (Adapted from Farooq et al., 2009 and Tuong et al., 2005, with modification).

in one location to irrigate new land in another, and this can only occur in water-limited environments when land area is not the limiting factor (Gaydon et al., 2012a). The other possibility is that irrigation water saved during one crop could be used to better irrigate a second (or third) crop as part of a rotation on the same field, providing any of those subsequent crops were originally water-limited. These crops could be additional rice crops, or else non-flooded crops like wheat, maize or legumes.

So in conclusion, the Punjab example illustrates that water-savings at a field scale may not necessarily flow on to water savings at higher scales. The most scale-independent improvements in irrigation efficiency are those which reduce ET (Humphreys et al., 2010), although in certain hydrogeological circumstances reduction in deep drainage and lateral water losses may also be true savings. It is therefore important to understand the hydrogeology of the region of interest, as this will define the criteria on which the effectiveness of irrigation water savings can be judged. Also, water savings and increased water productivity at the field scale are not a means to produce more rice, unless the water saved can be diverted to additional irrigated land or crops. This implies that water savings are not a desirable aim in land-limited production environments (Gaydon et al., 2012a), and that additional cropping land (or other water-limited crops in the same rotation) are pre-requisites to realise the benefits of any water-savings achieved (Bouman, 2007a).

In the following sections we will review the current status and knowledge for a number of water-saving irrigation techniques (Fig. 1), and assess them in view of these perspectives. Figure 1 illustrates these practices on a scale of increasing water availability for the growing rice crop. At the low end of the spectrum, traditional upland systems are rainfed, often on poor soils, rarely ponded and rice yields are generally low. Increased water productivity in these systems may be achieved via agronomic practices such as residue retention to reduce soil evaporation, but since they are not irrigated, they will not be considered further in this chapter. Aerobic rice systems are irrigated but not ponded, with a focus on water economy. Moving further up the spectrum towards fully ponded lowland rice systems (which may be rainfed or irrigated), we encounter interim irrigated practices such as alternate wetting and drying (AWD) and the system of rice intensification (SRI). These also aim for water savings over fully ponded systems, but seek higher yields than aerobic systems, with less water savings. More complete descriptions of these systems, together with assessment of benefits and risks, are discussed further in the following sections.

4 Aerobic rice as a water-saving measure

4.1 Description

Aerobic rice is a term introduced by the International Rice Research Institute (IRRI) for high-yielding rice grown under non-flooded conditions in non-puddled and unsaturated (aerobic) soil. The crop is responsive to nutrient supply, can be rainfed or irrigated, and tolerates (occasional) flooding (Bouman and Tuong, 2001). On the sliding water-availability scale figure (Fig 1.) it can be seen that aerobic rice fits in between AWD and pure rainfed upland rice. In an aerobic rice system, the crop can be dry direct-seeded or transplanted and soils are kept aerobic throughout the growing season. Supplemental irrigation is applied as necessary (Nie et al., 2012). A key component of success in aerobic rice system is selection of appropriate cultivars (Wang et al., 2002), as root exploration and drought tolerance are key selection criteria. For example under declining soil moisture (measured as a % of the available soil water) most crops maintain transpiration up to around 30% of available soil water after which the rate declines (Loomis and Connor, 1992); in rice, there appears to be a linear decline from around 70% of available soil water (Lilley and Fukai, 1994). Recent varieties bred specifically for aerobic conditions are targeted at approaching the behaviour of other non-rice crops (wheat, maize, etc.).

There are a range of ways in which irrigation water may be applied in aerobic rice systems. These include sprinkler irrigation (Kahlown et al., 2007), furrow irrigation on beds (Choudhury et al., 2009) and drip and subsurface irrigation (Parthasarathi et al., 2014).

It has been stated that weeds are the most severe constraints to widespread adoption of aerobic rice (Rao et al., 2007; Jabran and Chauhan, 2014); however, rice yield decline reported in a number of studies after years of continuous aerobic rice cultivation clearly represents another severe constraint which needs to be addressed (Kreye et al., 2009).

4.2 Benefits

Growing rice under continuously unsaturated soil conditions can maximise water-use efficiency and minimise both labour requirements and greenhouse-gas emissions (Kato et al., 2014). In temperate environments, aerobic rice culture regularly produces grain yields greater than 9 t ha^{-1} (e.g., in yields in the United States under centre-pivot sprinklers reach 10 t ha^{-1}, with up to 11.4 t ha^{-1} achieved in central Japan). However, yields remain at less than 8 t ha^{-1} in the tropics (Kato et al., 2014).

The harshness (heat, evaporative demand) of the environment is pertinent to how well rice responds to aerobic irrigation, with large yield reductions in sprinkler irrigated rice of 35–70% reported in temperate Australia (Muirhead et al., 1989) compared with conventional flooded yields of 10$^+$ t ha^{-1}. Associated water savings were similar percentages resulting in negligible changes to WP; however, it must be noted that the varieties used in the experiment were not adapted to aerobic conditions and were typical local lowland rice varieties (Humphreys et al., 1989). In Brazil, aerobic rice cultivars with high grain yields of 5–7 t ha^{-1} have been developed (Castañeda et al., 2002). While in northern China, the grain yields of 8 t ha^{-1} and even higher have been achieved using high-yielding aerobic rice cultivars under appropriate management practices (Wang et al., 2002).

By reducing water use during land preparation and limiting seepage, percolation, and evaporation, aerobic rice had 51% lower total water use and 32–88% higher water productivity, expressed as gram of grain per kilogram of water, than flooded rice (Bouman et al., 2005). Castañeda et al. (2003) reported that rice yields under aerobic conditions were 2.4–4.4 t ha^{-1} (14–40% lower than under flooded conditions), but that water use decreased relatively more than yield, resulting in irrigated water productivity increases by 20–40% (in one case even 80%) over that under flooded conditions. Water savings averaged 73% of irrigation water for land preparation and 56% during the crop growth period, compared with conventional flooded production (Castañeda et al., 2003).

The above studies were all conducted with the same cultivars used for flooded and aerobic cultivation. When specific aerobic cultivars are employed under aerobic conditions and compared with specific lowland cultivars under lowland conditions, the reported WP gains are even higher. The WUE of the aerobic varieties under aerobic conditions can be 164–188% higher than that of a lowland cultivated rice varieties (Farooq et al., 2009).

The use of surface mulches in aerobic rice systems has been noted to increase irrigation water productivity (Qin et al., 2006), significantly increasing the leaf area per plant, main root length, tap root length and root dry weight per plant of crop, in addition to reducing soil evaporation losses. In an experiment in China comparing flooded rice irrigation with aerobic rice under different surface mulching scenarios, it was found that WP was increased in the aerobic rice system by 307–321% under surface mulching, but only by 98–138% under the no-mulch treatments (Zhang et al., 2008). Obviously there are socio-economic aspects to the use of mulch in small-holder farming systems which must be taken into consideration (requirements for livestock feed, etc.); however the biophysical benefits of mulching in aerobic rice systems, from a grain production and irrigation water perspective, appear clear.

There is also evidence that the intermittent irrigation characteristic of aerobic (or AWD) rice systems may be a useful tool to fight the build-up of arsenic in flooded rice cultivation of boro rice production in Bangladesh (Roberts et al., 2010). This appears to be largely driven by the disruption of saturated soil redox conditions under aerobic or AWD management.

Labour use is also saved in aerobic rice because more labour is required for land preparation such as puddling, transplanting and irrigation activities in flooded rice (Wang et al., 2002; Farooq et al., 2009).

Less methane emissions are expected under aerobic than under flooded conditions, but with higher nitrous oxide emissions (Wassmann et al., 2000). Weller et al. (2016) considered the combined global warming potential (GWP) due to both CH_4 and N_2O and reported the GWP of conventional flooded paddy rice as 3868 ± 2137 kg CO_2-eq ha^{-1} $season^{-1}$, compared with 2029 ± 1050 kg CO_2-eq ha^{-1} $season^{-1}$ for an aerobic rice system, over a 4-year trial. This represents a reduction in global warming potential of 47% for aerobic rice irrigation. However, there are two confounding elements in this evaluation: (i) when this is scaled according to grain yield the reduction was less, but (ii) the aerobic rice system also comprised one fully-flooded (wet season) crop per year and one aerobic crop. The results are also significantly affected by soil type and climate. At this stage it is fair to say there is little published evidence that aerobic rice systems present better GWP outcomes than AWD systems; however, enhanced performance compared with conventionally flooded systems has been clearly demonstrated.

4.3 Risks

With continuous aerobic rice, there have been concerns reported regarding the so-called 'soil sickness'. Yield decline resulting from continuous cropping of aerobic rice has been identified as a major constraint to the widespread adoption of aerobic rice technology (Nie et al., 2012). The decline also appears to be greater in the dry season than in the wet (Peng et al., 2006); however the magnitude of yield decline after continuous cropping of aerobic rice depends strongly on the number of seasons that aerobic rice was continuously cropped, soil properties, climates, rice cultivars, and management practices (Kreye et al., 2009).

It has been suggested that growth-limiting factors other than water or N affected the crop in these situations, from the stage of panicle initiation onwards. Root knot nematodes and micronutrients have been nominated as possible major constraints to crop development in these continuous aerobic rice situations (Kreye et al., 2009). Strategies to overcome continuous cropping obstacle of aerobic rice have been developed. These strategies include crop rotation, N management practice, soil acidification, and cultivar improvement (Nie et al., 2012). The problem is not unknown in other cropping systems without appropriate measures including rotations and crop-breaks (Barker and Koenning, 1998; Thompson et al., 2012; Grabau and Chen, 2016), and breaking up the lifecycle of the pest (for example nematodes) is the aim. A detailed review of the literature on options for managing yield decline in aerobic rice systems is found in Nie et al. (2012).

Loss of soil organic carbon reserves has also been identified as a risk in aerobically irrigated rice systems. The introduction of aerobic phases in lowland rice systems has been found to result in decreased organic C and total N due to low C sequestration in soils (Witt et al., 2000). A recent study by Kraus et al. (2016) indicated that under land management change which sees flooded paddy rice replaced by an aerobic or upland crop, approx. 2.8–3.4 t C ha^{-1} yr^{-1} residue incorporation after harvest is required to achieve stable soil organic carbon (SOC) stocks in the new system. This suggests that aerobic rice irrigation systems may need to include some principles of conservation agriculture (e.g. residue retention) to maintain long-term productivity.

Aerobic rice may also be a risky practice in saline-prone environments, as the reduced water application will likely result in reduced flushing of salts below the root zone, increasing the potential for greater build-up of salts and consequent crop salt stress.

5 Alternate wetting and drying (AWD) as a water-saving measure

5.1 Description

An established irrigation water-saving technology for rice is *alternate wetting and drying* (AWD) developed earlier but document by IRRI in the 1970s (Bouman and Tuong, 2001; Belder et al., 2004; Sandhu et al., 1980). This technology is based on the fact that high yields in rice can be achieved without continuous flooding. Once the crop is established, the pond depth is allowed to fall to a threshold depth below the surface of the soil surface for a certain period before the next irrigation is applied. In 2001, the IRRI developed a set of simplified guidelines for AWD technology using a field water tube as the tool to monitor the water level below the soil surface. Irrigation is applied (to a depth of around 5 cm) when the perched water table falls to 15 cm below the soil surface. The field water tube (Fig. 2a and b) is a perforated tube that can be fabricated with materials such as polyvinyl chloride (PVC) pipe, bamboo, plastic bottles, or even tin cans, with diameter ranging from 10 to 20 cm (Bouman et al., 2007b).The threshold of 15 cm is called 'safe AWD', as this will not cause any significant yield decline. In more technical terms, irrigation is applied when soil water tension increases to 10 kPa at a depth of 15 cm (Sudhir-Yadav et al., 2011a). Safe AWD is now recommended to farmers in several Asian countries, including Vietnam, India, Bangladesh and the Philippines (Lampayan et al., 2015; Palis et al., 2014).

It has been demonstrated over many years now in numerous environments that rice can extract adequate water from the soil around roots, even in the absence of standing water in the field (Lampayan et al., 2015). In some circumstances there are constraints on which phases of the crop growth can be irrigated in this way (see following case study from Australia's Riverina (Dunn and Gaydon, 2011)); however, in many environments, the majority of crop growth period can be irrigated using AWD, with the sensitive period around flowering exempted (flooding of 5cm applied for 1–2 weeks).

5.2 Benefits

Whereas practically all studies report significant decreases in water use (mm/ha) with AWD in comparison with continuous flooding, some studies have actually reported an increase in rice grain yield (Zhang et al., 2008; Won et al., 2005; Yang et al.,2007), whereas others report a decrease in rice grain yield (Tabbal et al., 2002; Belder et al., 2004). In 2001 Bouman and Tuong asserted that AWD would always result in some degree of yield decline due to the inevitability of incurring some level of rice crop water stress; however, subsequent studies have indicated this is not always the case.

In a 4-year Chinese experiment, Feng et al. (2007) demonstrated that rice yields under flooded conditions were around 8000 kg ha^{-1} with 900 mm total (rain, irrigation) water input. Irrigation water savings of 40–70% were achieved without any yield loss by applying AWD. In a different region of China, Belder et al. (2004) reported that the AWD technology can reduce field water use by 15–20% without significantly affecting yield. In Hubei province, another experiment showed that AWD reduced irrigation water without significantly impacting grain yields and increased the mean water productivity by 16.9% compared with continuously flooded irrigation (Tan et al., 2013). Sudhir-Yadav et al. (2011a,b) demonstrated the feasibility of reducing irrigation amount (30–50%) while maintaining yield by replacing puddled transplanted rice (PTR) with direct-seeded rice (DSR) using AWD in the Indian Punjab, provided that soil tension is kept lower than 20 kPa at 20 cm depth. However, in the Philippines, in an area with a relatively deep groundwater table, intermittent irrigation every 8 or 10 days reduced the rice yield on average by 25% over that of the control and the average water input by 60% that of the control. Water productivity, however, still nearly doubled the continuously flooded value (Tabbal et al., 2002). But with 20 days of continuous drought, yields declined so much that even water productivities declined to 82% of the value under continuous flooding. In experiments in Yangzhou, Eastern China, the grain yield was increased by 9.3–9.5% under AWD with moderate soil drying between anthesis and maturity (limit of -25 kPa at 15–20cm depth), while it was reduced by 7.5–7.8% under a more severe form of AWD over the same period (limit of -50kpa), compared with that under conventional flooding. Water applied to the moderate AWD treatment was 44% and to the severe AWD treatment was 25% of the amount applied to the conventionally flooded treatment. The moderate AWD also significantly improved milling, appearance and cooking qualities, while the severe AWD decreased these qualities, once again in comparison with continuously flooded production. The authors concluded that a moderate wetting drying regime during the grain-filling phase holds great promise to increase both yield quantity and quality and also to save water (Zhang et al., 2008). The authors also suggested that soil drying to -25 kPa is notably beneficial during the grain-filling period, but remarked that during their experiments relatively cool conditions prevailed during this crop growth phase and questioned whether the same results would be achieved under warmer tropical conditions.

So this raises the hypothesis that climate may explain discrepancies, although it is likely more complex than that. In summary, the discrepancies in the water savings/yield trade-offs with AWD are probably due to variations in any number of things: soil characteristics, depth to water-table and other local hydrological aspects, timing of the applied water stress, and also varietal differences. It is also possible that different climates during key crop stages are influential. Although some varieties are likely to perform better under AWD than others, it appears that benefits are available to all. In Hubei, China 2009–10, Super Hybrid rice developed for high-yielding lowland production was compared under continuous flooding (CF) and AWD. AWD saved 24% and 38% irrigation water compared with CF in 2009 and 2010, respectively. There was insignificant difference in grain yield between AWD and CF. These results suggest that high-yielding varieties developed for the continuously flood-irrigated rice system could still produce high yield under safe AWD experienced in this study. 'Super' hybrid rice varieties do not necessarily require more water input to produce high grain yield. (Yao et al., 2012). It is also likely that varietal traits including drought tolerance and deep rooting will be advantageous in AWD systems and can act to reduce any observed yield losses resulting from water stress.

It seems clear that AWD improves WP universally when applied within recommended guidelines; however as illustrated above, the specific impacts on and trade-offs between irrigation water use and rice grain yields will vary. Whether AWD will solve water shortage problems is also not clear-cut. Bouman et al. (2007) showed that most of the irrigation water savings from AWD are caused by reduced deep drainage (or percolation) rates. This will reduce groundwater recharge and may lead to decreased opportunities for groundwater irrigation in some regions (e.g. NW India, Humphreys et al., 2010). In Punjab, Aurora et al. (2006) showed 350 mm of water savings from AWD in PTR (25–30% of continuously flooded water use), but ET was only reduced by about 30 mm. From this perspective of water savings, AWD is unlikely to be a key solution to the crisis of falling water tables in the North-West India, where technologies to decrease in ET are the main objective (Humphreys et al., 2010). However, in areas where deep percolation losses are effectively lost from the farmer's irrigation system, AWD water savings may be a real solution to any water-shortage problems.

In addition to water saving advantages, the AWD practice was reported to reduce methane (CH_4) emissions from rice fields (Hosen, 2007) due primarily to reduced time in a flooded state under reduced soil conditions. Among many agricultural practices, water management has been recognised as one of the most promising approaches to reduce CH_4 emission (Wassmann et al., 2000; Li et al., 2002). Midseason drainage is a common practice used by farmers to control the unproductive tillers. Field studies have shown that midseason drainage reduces CH_4 emissions by above 10% (Wassmann et al., 2000). In Chinese experiments, the CH_4 emissions under AWD_{15} (15cm threshold for irrigation) and AWD_{30} were more than 30% and 60% lower than under CF, respectively (Liang et al., 2016). However reductions in CH_4 are only part of the emissions picture when changing from CF to AWD – the increased drying and re-wetting of soil associated with AWD irrigation also increases nitrification (on soil drying) and denitrification (on soil wetting), resulting in the release of N_2O and other gaseous N products, which have higher global warming potential per mass than CH_4.

Carbon Dioxide equivalents are a universally accepted method of integrating the net effect of gaseous C and N products on global warming potential (Robertson et al., 2000). Several studies are now available which explore the global warming potential of different rice water management systems, and the literature reveals a common conclusion (Linquist et al., 2015). For example, carbon dioxide equivalents of CH_4 and N_2O emissions from paddy fields under AWD irrigation were found to be 788 kg CO_2 ha^{-1}, a reduction of 61.4% compared with those from conventional flooded irrigation (Yang et al., 2012). In summary, N emission are increased, but C emissions are decreased under AWD water management, with the net impact being reduced carbon dioxide equivalents. It can therefore be concluded that AWD is an effective technique for mitigating the global warming potential of irrigated rice production.

There is evidence available that the investment to develop and disseminate the AWD technology has had a high rate of return, with benefit-cost ratio of 7:1. The evidence of economic benefits at the farm level when aggregated will more than compensate for the total research investments made to develop and disseminate the AWD technology (Lampayan et al., 2015).

5.3 Risks

Although AWD can substantially reduce irrigation water use at the field scale, there is the risk that excessive AWD drying in puddled clay-based soils can inadvertently create the opposite outcome due to development of deep cracks extending below the plow-pan, which actually increase total crop water use from rapid bypass flow upon re-wetting, and before the cracks self-seal (Bouman and Tuong, 2001). In Hubei province, China, the deep drainage under AWD was reduced by 15.3% in 2007 and 8.3% in 2008 compared to CF (Tan et al., 2103). However, the cumulative percolation of the first 5 days after irrigation in AWD plots was significantly larger than that in CFI plots. The authors concluded from this that bypass flow across the plow-pan, facilitated by the opening of cracks in the unstructured (puddled) soil during the drying phase, allows significantly increased water losses in the period before the cracks 'self-close'. This potentially decreases the water-saving effectiveness of AWD and increases the potential for NO_3-N loading to the groundwater.

Similar to aerobic rice irrigation, AWD may also be a risky practice in saline-prone environments, with increased potential for greater build-up of salts and consequent crop salt stress due to reduced leaching of salts (compared with continuously flooded systems).

6 Saturated soil culture (SSC) as a water-saving measure

6.1 Description

Saturated Soil Culture (SSC) is an irrigation practice which involves keeping the rice-growing soil saturated but not ponded throughout the growth period, thereby reducing the hydraulic head of the water, and in turn reducing the percolation rate and water lost to deep drainage. Evaporative losses are likely to be similar to conventional flooded practice (Borrell et al., 1997).

SSC is the typically nominated water management method in the System of Rice Intensification (SRI) (initially described in Stoop et al., 2002; Uphoff et al., 2002), at least up until panicle initiation, after which a very thin pond of 1–2 cms is recommended. Numerous authors have either debunked (Sheehy et al., 2004; McDonald et al., 2006) or supported (Hussain et al., 2004; Sinha et al., 2007) the remarkable claims of yield and productivity increases available through SRI; however for the purposes of this chapter, water savings available through employing SSC in comparison with conventional flooded practice are reported as positive. This section will not attempt to delve into issues relating to SRI grain production potential – only to examine potential water savings through implementation of SSC irrigation practices.

6.2 Benefits

From a large Indian, Philippine and Japanese dataset, Bouman and Tuong concluded that under saturated soil conditions water savings were on average 23% (±14%) with yield reductions of only 6% (±6%), compared with production through maintaining a 5–10 cm pond throughout crop growth. This was noted as the most promising option to save water and increase WP without decreasing land productivity too much. According to Bouman and Tuong (2001), although substantial increases in WP were possible when applied following 'safe AWD' protocols, AWD always results in reduced land productivity. Overall production increases were then only possible by following a philosophy of 'spreading the water' over wider cropping areas (Gaydon et al., 2012a), or else diverting saved water to another crop in the rotation on the same field.

In a two-season experiment in Northern Australia on a permeable soil, Borrell et al. (1997) found that there was no significant difference in rice yield or rice quality between SSC and conventional flooded practice; however, there was approximately 32% saving in irrigation water use from SSC in both seasons. Water productivities (defined as weight of grain produced per mm applied water) were increased by up to 26%.

In a Chinese investigation comparing SRI to conventional flooding, Zhao et al. (2010) reported that irrigation use was reduced by 57.2% through SRI methods. Since they also recorded significant gains in rice yield with SRI, an impressive increase in irrigated WP of 194.9% was achieved; however the latter stages of their SRI crops were irrigated in an AWD pattern every 3–7 days. The authors also noted that SRI may be worth further study, solely on the basis of possible water savings – irrespective of the sometimes fierce debate over the impact of SRI on grain yield potential. Thakur et al. (2016), in summarising the current state of information on SRI, reported that water savings are possible up to 40% using pure SSC irrigation methods (i.e. without AWD).

Although claims on the increasing water productivity of SRI-practices, including reduced water use, are found in many publications, there is a surprising lack of studies quantifying the magnitude of these changes under farmer conditions (Berkhout et al., 2015). Adusumilli et al. (2011) reported irrigation water savings of 52% for SRI over conventional practice, and Gathorne-Hardy et al. (2013) indicated a 55% water saving, for an increase in WP of 139%; however, all of these figure represent a mix of SSC (prior to panicle initiation) and AWD (till maturity).

In summary, the pure SSC culture studies report irrigation water-savings ranging between 9 and 40%. This range is most likely due to the range of soil permeability used in experiments. The literature also indicates that in combination with AWD, SSC has the potential to increase water-savings higher, up to 52–57.2%.

Reduced total global warming potential associated with SRI technology has been reported (1.1 kg CO_2 equivalents per kg paddy rice, compared with 2.8 for conventional flooded practice; Gathorne-Hardy et al., 2013). However, once again, the water management employed in this system is a mix of SSC and AWD.

6.3 Risks

There is a risk of increased rice floret sterility in cold weather, particularly relevant in temperate climates. SSC provides no thermal buffering protection to the rice crop from the water mass, such as can be achieved in fully ponded production (Williams and Angus, 1994). In these ponded systems, the developing rice floret may be submerged into warmer water to avoid low air temperatures. Authors also note increased weed control problems over conventional flooded practice which require additional inputs in terms of herbicide and/or labour (Borrell et al., 1997; Ly et al., 2012; Gathorn-Hardy et al., 2016).

7 Case study: water-saving irrigation in southeast Australia

In temperate southeast Australia, japonica rice is traditionally grown under fully irrigated conditions over a long growing period (close to six months, starting October and going through till April). Consequently grain-filling occurs over an extended time, and grain yields are among the highest in the world (12+ tonnes/ha). Since the mid-1980s the average water productivity (WP) of the total New South Wales (NSW) rice crop has effectively doubled (Humphreys et al., 2006). The breeding and introduction of high-yielding semi-dwarf cultivars, together with reduced field water use, have largely driven this improvement; however, several climate-specific adaptations in management practice have also been influential. The climate constraints in this region are significantly different from much of the tropical rice-growing world – the average rice crop in southeast Australia requires around 1200 mm of water throughout its growing cycle, yet on average only about 350 mm of that comes from rainfall (Humphreys et al., 2006). The rest is supplied via irrigation from either diverted surface water or groundwater pumping. During the growing period the crop experiences hot, dry weather with daytime temperatures often rising above 40°C and night-time temperatures falling below 15°C on a regular basis. Historically, the risk of the crop being exposed to night-time temperatures less than the damaging 17°C threshold during the sensitive microspore period is around 40% – a very high production risk due to low-temperature floret sterility (Williams and Angus, 1994) with the potential for dramatic yield reductions. The application of deep water (20–25 cm) during this period is a recommended method adopted by farmers to protect the pollen against low temperature (Humphreys et al., 2006). Australian rice growers are regarded as rapid adopters of improved technology (McDonald, 1994; Clampett et al., 2000), and the deep-water flooding technology at microspore has had an almost universal uptake by industry.

Historically, water shortages were never a problem for Australian rice producers, receiving at least 100% of their requirements every season from as far back as 1912 (at the commencement of the irrigation schemes) right up until the mid-1990s (McIntyre et al., 2011; Gaydon et al., 2012a). At this point, a mix of political and climatic factors led to unprecedented restrictions in production due to low water availability and extreme variability of supply in the intervening period till the present (Gaydon et al., 2012a). This has driven a growing interest in water-saving technology.

There was thought to be little opportunity for increasing water productivity from the current flooded rice-growing practice (Humphreys et al., 2006), so farmers and researchers looked to AWD and realised that one opportunity to save water may be through reducing the length of time the crop is flooded. The period when rice must be flooded in deep water for cold-temperature protection during the microspore period presented a clear constraint (Williams and Angus, 1994). Also, flooding is required during the subsequent reproductive and grain-filling period to meet the crop water requirements during the very high crop growth rates (250–300 kg/ha/d) that occur between panicle initiation (PI) and flowering. The high potential evapotranspiration demand in this very dry environment over this period could result in crop water deficit stress if a non-ponded culture was used (Humphreys et al., 2005) and result in significant yield losses. The early phase of crop growth from establishment to the time when the crop reaches full canopy cover offered particular potential for reducing water use using AWD. During this period the plant canopy is small and evaporation from the free water surface constitutes up 40% of total evaporation loss in continuously flooded rice (Simpson et al., 1992).

In the early 1980s AWD was first trialled in Australia. Heenan and Thompson (1984) recorded water savings of 23% through AWD every seven days prior to the establishment of continuous flooding at PI, achieving equivalent grain yields to conventionally flooded practice. Subsequent studies on different sites also demonstrated an increase in water productivity, with AWD ranging from 0.06 to 0.23 kg/m^3 compared to the normal flooding regime (Thompson and Griffin, 2006). However, all these early studies with AWD in Australia involved regular irrigation intervals with little moisture stress on the crop during the unflooded period. It was thought that greater water savings may be achieved with less frequent irrigations, but the optimal level of moisture stress between irrigation was unknown.

Dunn and Gaydon (2011) therefore conducted experiments investigating water savings from a range of treatments, flush irrigated at different cumulative evapotranspiration frequencies prior to the start of continuous flooding, which was delayed until just prior to panicle initiation. They demonstrated increased input water productivity from higher levels of imposed crop water stress during the initial non-flooded period. Irrigation intervals at 160 mm cumulative ET (equivalent to around 10 day AWD intervals) significantly improved input water productivity above that of the conventional drill sown treatment (by 17%). Irrigating at 80 mm intervals resulted in a significant but lesser improvement (9%). Notable from these experiments were the vigour with which the crops in the highly stressed AWD treatments bounced back once they received full flooding at PI – it was difficult to visually distinguish them from the conventional treatments during grain filling, and yield effects were minimal. However delaying the application of continuous flooding extended the period of crop growth and the authors concluded that it may be desirable to move sowing forward 7–10 days so that pollen microspore could still occur at the safest time in regard to cold temperature damage. A subsequent modelling study extending this experimental work and focusing on long-term risk found that establishing the crop up to a month earlier with the harsher AWD imposition could minimise the low-temperature risk and maximise system WP, but indicated that projected future climate changes by 2030 could see the optimum crop establishment time reverting to the original optimum establishment time (Gaydon et al., 2012b).

This case study is an example of how environmental constraints can limit the options for applying water-savings irrigation technology, but how targeted research combining field trials and modelling can allow the development of smart technology which takes such environmental constraints into account. Locally targeted research to fully understand the interactions between crop, management and environmental constraints and to subsequently use this understanding to optimise practical advice to farmers represents the greatest challenge to researchers in the field of irrigation water-savings. Participatory engagement with farmers allows researchers to gain insights into system constraints which may limit farmer options. Once these types of understandings are obtained, field experimentation in combination with cropping systems modelling offers researchers insights into systems performance and risks over larger timescales, and indeed in potential future climates. This is likely to be of growing importance to future researchers in the field of irrigation and crop production, as we move into climate conditions for which past experiences can yield us limited insights.

8 Future trends and conclusion

The call to 'grow more rice with less water' (Guerra et al., 1998) clearly requires more efficient use of water in irrigation and increased rice WP through advances in irrigation techniques; however this review has revealed that such a goal is more complex than it first seems. Water-saving irrigation increases water productivity, but in rice generally results in decreased yield (Tuong et al., 2005). There are some examples of concurrent water savings and yield increases (Liang et al., 2016), but in general water-savings do not produce more rice with less water in the same field. Field-level water productivity and yield can best be increased concurrently by water-saving irrigation plus raising crop yields through varietal improvement or better agronomy (fertiliser management, weed and pest control, etc.). Total rice production increases from water savings alone can occur by using the water saved in one locality to irrigate new land in another, or by diverting saved water onto another sequential water-limited crop (either another rice crop or non-flooded crop) in rotation on the same field. If this is not done or is not possible, a strategy of saving water at the field level potentially threatens total rice production at large (Bouman and Tuong, 2001). This implies that water savings alone are not a desirable aim for land-limited production environments, but require additional available land and crops in which to expand production to realise benefits (Gaydon et al., 2012a; Bouman, 2007a).

Water savings at a field scale may not necessarily flow on to water savings at higher scales. The most scale-independent improvements in irrigation efficiency are those which reduce ET (Humphreys et al., 2010), although in certain hydrogeological circumstances reduction in deep drainage and lateral water losses may also be true savings. It is therefore important to understand the hydrogeology of the region of interest, as this will define the criteria on which the effectiveness of irrigation water savings can be judged.

A review of current literature has revealed that the AWD rice irrigation technology is capable of achieving 15–70% water savings over conventional flooded irrigation practice, without any significant yield reductions. The savings strongly correlate with soil percolation rate rather than evaporative demand, hence varying greatly geographically as indicated by this large per cent savings range. This 'Safe AWD' has been widely extended within the rice-growing world with a demonstrably high return on research investment. For some environment-variety combinations, significant increases in crop yields have even been demonstrated, resulting in even greater WP gains. Saturated soil culture (SSC) irrigation such as those demonstrated by SRI spans the gap between AWD and conventional practice, but seems more limited in its application due to more intense management requirements and reduced water savings.

In regions where water availability is lower, aerobic rice systems may be the preferred irrigation practice with greater yield reductions than AWD, but with potential for even larger irrigation water use savings. Whether this necessarily leads to greater WP than AWD systems may depend on variety, soil, and environmental conditions. But to realise the advantages of these aerobic (or AWD) WP gains, there must be extensive cropping lands available in which to 'spread the saved water' and increase cropping area, or else to more intensively irrigate another previously water-stressed crop in the rotation. Aerobic rice is not the system of choice for land-limited cropping systems without significant water shortages. Also, it is impossible to escape from the fact that rice is still a drought-sensitive crop, which prefers high-moisture content soils. There are limits for stressing rice crops in aerobic irrigation beyond which WP plummets due to great reductions in grain yields. In many cases where irrigation water is more limiting, it may be more productive to focus on other dry-season crops like wheat, maize, legumes.

The basic biophysical advantages of rice irrigation water-savings technologies detailed in this chapter are now quite well understood from experimental studies under a range of conditions and environments. However, less is known about potential savings extrapolated to wider geographical areas incorporating a range of soil-types, water availabilities and climates, or the impact of regional socio-economic constraints when these techniques are employed within cropping systems management practices like Conservation Agriculture (Hobbs, 2007). Hence a future trend in irrigation research for rice-based systems will surely be integrating existing knowledge using tools like crop models, water-resources and climate models, together with global information systems (GIS) to draw useful conclusions at a regional or policy scale regarding initiatives to increase water productivity in irrigation. A hydrological characterisation and mapping of Asia's rice area is needed to assess the extent and magnitude of potential water savings, because the gains vary with soil type and climate (Belder et al., 2004).

Cropping systems philosophies like Conservation Agriculture also raise new questions about proven rice irrigation practices. For example, what is the potential to use AWD with unpuddled soils (less cracking during drying phase than puddled soils, due to maintenance of soil structure) to reduce the risk of bypass flow transporting NO_3-N to the ground water (Tan et al., 2013)? What are greenhouse gas emission and soil organic carbon implications of land management change options? Some studies have started to investigate these broader cropping system issues (e.g. Kraus et al., 2016; Kadiyala et al., 2014) and these are regarded a pertinent future trends in these irrigated systems.

The *crop-scale* questions explored by experiments of the past will now morph into *systems-scale questions*. For example, how best to irrigate rice in rotation with other non-flooded crops and pastures? To which on-farm enterprises are the water resources best allocated? Is it more profitable for the farmer to optimise his rotational system, rather than optimise the performance of any one crop? Should irrigation water savings from one crop be employed on another subsequent crop in the same field, or should the area of cultivation be expanded? Is limited irrigation water best used on a rice crop, or on another type of crop (wheat, maize etc.)? Which option gives the best outcomes for farmer profitability, or the best outcomes for regional food security?

As food production pressure increases in coming years, there will be an increasing pressure to produce food from marginal lands and harsh environments. Questions for irrigation scientists in rice-based systems may include, 'How are we best to employ brackish irrigation water to expand available supply and hence crop-able area?' 'What is the best strategy to mix available fresh water supplies with other water supplies of lesser quality?' and 'When should they be used on the crop?'

Climate change adaptation will obviously figure strongly in future irrigation science research trends. What will be the future climate impacts on irrigation water requirements? On irrigation water availability? On irrigation water quality? How will these affect the crop yields, water savings and water productivities possible via AWD and aerobic technologies, and how will they increase the risk or variability of returns?

Cropping systems and water-resources models are useful tools to explore future scenarios, provided they are well parameterised, calibrated and validated locally for the region of interest. They can be used to extend what we learn from a limited number of years in field experiments and to give insights into system performance and risk over much longer timescales and in different climatic conditions. For many of these questions raised for the future of irrigation in rice system, well-validated cropping systems models, when used judiciously together with field experimentation, will allow irrigation scientists of the future to explore the effects of future conditions over broad expanses of the rice-growing world and evaluate performance of different strategies and technologies. It is difficult to see how traditional field-based agronomy and irrigation studies can yield the required information on how best to irrigate rice-based systems into the future without the concurrent use of cropping systems and water-resources models. The field-experimentation/modelling nexus is likely to be a growing trend in rice irrigation research as we prepare to feed humanity in the coming centuries.

9 Where to look for further information

The following references provide enhanced detail on many of the concepts raised in this chapter. Additionally, the reader may gain valuable information from the IRRI website (http://irri.org/resources) and a range of freely available resource publications therein. The International Water Management Institute (IWMI) also has publicly accessible resources related to water management and water savings options (http://www.iwmi.cgiar.org/publications/latest/).

10 References

Adusumilli, R. and Bhagya Laxmi, S., 2011. Potential of the system of rice intensification for systemic improvement in rice production and water use: the case of Andhra Pradesh, India. *Paddy and Water Environment* 9, 89–97.
Alberto, M. C. R., Wassmann, R., Hirano, T., Miyata, A., Kumar, A., Padre, A. and Amante, M., 2009. CO 2/heat fluxes in rice fields: comparative assessment of flooded and non-flooded fields in the Philippines. *Agricultural and Forest Meteorology* 149(10), 1737–50.
Anthofer, J., 2004. *The Potential of the System of Rice Intensification (Sri) For Poverty Reduction in Cambodia*. Conference on International Agricultural Research for Development, Berlin, 5–7 October.
Arora, V. K., 2006. Application of a rice growth and water balance model in an irrigated semi-arid subtropical environment. *Agricultural Water Management* 83, 51–7.
Barker, K. R. and Koenning, S. R., 1998. Developing sustainable systems for nematode management. *Annual Review of Phytopathology* 36(1), 165–205.
Barker, R., Dawe, D., Tuong, T. P., Bhuiyan, S. I. and Guerra, L. C. (1998). The outlook for water resources in the year 2020: Challenges for research on water management in rice production. In 'Assessment and Orientation Towards the 21st Century', *Proceedings of 19th Session of the International Rice Commission*, 7–9 September 1998, FAO, pp. 96–109.
Belder, P., Bouman, B. A. M., Cabangon, R., Guoan, L., Quilang, E. J. P., Li, Y., Spiertz, J. H. J. and Tuong, T. P. 2004. Effect of water-saving irrigation on rice yield and water use in typical lowland conditions in Asia. *Agricultural Water Management* 65,193–210.

Berkhout, E., Glover, D. and Kuyvenhoven, A., 2015. On-farm impact of the System of Rice Intensification (SRI): Evidence and knowledge gaps. *Agricultural Systems* 132, 157–66.

Borlaug, N., 2007. Feeding a hungry world. *Science* 318(5849), 359.

Borrell, A., Garside, A. and Fukai, S., 1997. Improving efficiency of water use for irrigated rice in a semi-arid tropical environment. *Field Crops Research* 52(3), 231–48.

Bouman, B. A. M. and Tuong, T. P., 2001. Field water management to save water and increase its productivity in irrigated lowland rice. *Agricultural Water Management* 49(1), 11–30.

Bouman, B. A. M., 2007. A conceptual framework for the improvement of crop water productivity at different spatial scales. *Agricultural Systems* 93(1), 43–60.

Bouman, B. A. M., Lampayan, R. M. and Tuong, T. P., 2007b. *Water Management in Irrigated Rice: Coping with Water Scarcity*. International Rice Research Institute, Los Baños, Philippines, p. 54.

Bouman, B. A. M., 2013. Bas Bouman's blog – Global Rice Science Partnership, IRRI http://irri.org/blogs/bas-bouman-s-blog-global-rice-science-partnership/does-rice-really-use-too-much-water (accessed 29 April 2016).

Castañeda, A. R., Bouman, B. A. M., Peng, S. and Visperas, R. M., 2002. The potential of aerobic rice to reduce water use in water-scarce irrigated lowlands in the tropics: opportunities and challenges. In Bouman, B. A. M., Hengsdijk, H., Hardy, B., Bindraban, P. S., Tuong, T. P. and Ladha, J. K. (Eds), *Water-Wise Rice Production*, 8–11 April 2002, International Rice Research Institute, Los Baños, Philippines, pp. 165–76.

Castañeda, A. R., Bouman, B. A. M., Peng, S. and Visperas, R. M., 2003. The potential of aerobic rice to reduce water use in water-scarce irrigated lowlands in the tropics. In Bouman, B. A. M., Hengsdijk, H., Hardy, B., Bindraban, P. S., Tuong, T. P. and Ladha, J. K. (Eds), *Water-Wise Rice Production*, 8–11 April 2002 at IRRI Headquarters in Los Baños, Philippines, International Rice Research Institute, Los Baños, Philippines.

Choudhury, B. U., Bouman, B. A. M. and Singh, A. K., 2007. Yield and water productivity of rice–wheat on raised beds at New Delhi, India. *Field Crops Research* 100(2), 229–39.

Clampett, W. S., Williams, R. L. and Lacy, J. M., 2000. Major achievements in closing yield gaps of rice between research and farmers in Australia. In *Proceedings of the Expert Consultation on Yield Gap and Productivity Decline in Rice Production*, 5–7 September 2000. FAO, Rome, Italy, pp. 411–28.

Devendra, C. and Thomas, D., 2002. Smallholder farming systems in Asia. *Agricultural Systems* 71(1–2), 17–25.

Dunn, B. W. and Gaydon, D. S., 2011. Rice growth, yield and water productivity responses to irrigation scheduling prior to the delayed application of continuous flooding in south-east Australia. *Agricultural Water Management* 98(12), 1799–807.

Farooq, M., Kobayashi, N., Wahid, A., Ito, O. and Basra, S. M., 2009. Strategies for producing more rice with less water. *Advances in Agronomy* 101, 351–88.

Gathorne-Hardy, A., Reddy, D. N. and Harriss-White, B., 2013. A Life Cycle Assessment (LCA) of greenhouse gas emissions from SRI and flooded rice production in SE India. *Taiwan Journal of Water Conservancy* 61, 120–5.

Gathorne-Hardy, A., Reddy, D. N., Venkatanarayana, M. and Harriss-White, B., 2016. System of Rice Intensification provides environmental and economic gains but at the expense of social sustainability – A multidisciplinary analysis in India. *Agricultural Systems* 143, 159–68.

Gaydon, D. S., Meinke, H. and Rodriguez, D., 2012a. The best farm-level irrigation strategy changes seasonally with fluctuating water availability. *Agricultural Water Management* 103, 33–42.

Gaydon, D. S., Meinke, H. and Dunn, B. W., 2012b. Maximizing water productivity through model-aided design of delayed continuous flood irrigation in Australian rice production, Chapter 7 in Gaydon, D. S., 2012. Living with Less Water: Development of viable adaption options for Riverina irrigators, Ph.D thesis, Wageningen University, Wageningen, the Netherlands, p. 225. ISBN: 978-94-6173-232-3. http://library.wur.nl/WebQuery/wda/1987914.

Godfray, H. C. J., Beddington, J. R., Crute, I. R., Haddad, L., Lawrence, D., Muir, J. F., Pretty, J., Robinson, S., Thomas, S. M. and Toulmin, C., 2010. Food security: The challenge of feeding 9 billion people. *Science* 327, 812–18.

Grabau, Z. J. and Chen, S., 2016. Influence of long-term corn–soybean crop sequences on soil ecology as indicated by the nematode community. *Applied Soil Ecology* 100, 172–85.

GRiSP (Global Rice Science Partnership). 2013. *Rice Almanac*, 4th edition, Los Baños (Philippines), International Rice Research Institute, p. 283.

Guerra, L. C., Bhuiyan, S. I., Tuong, T. P. and Barker, R., 1998. *Producing More Rice With Less Water From Irrigated Systems*. SWIM Paper 5. IWMI/IRRI, Colombo, Sri Lanka, p. 24.

Hobbs, P. R., Sayre, K. and Gupta, R., 2008. The role of conservation agriculture in sustainable agriculture. *Philosophical Transactions of the Royal Society of London B: Biological Sciences*, 363(1491), 543–55.

Heenan, D. P. and Thompson, J. A., 1984. Growth, grain yield and water use of rice grown under restricted water supply in New South Wales. *Australian Journal of Experimental Animal Husbandry* 24, 104–9.

Hobbs, P. R., 2007. Conservation agriculture: what is it and why is it important for future sustainable food production? *Journal of Agricultural Science* 145(2), 127–37.

Hosen, Y., 2007. Whether water saving reduces the global warming potential of irrigated paddies. *Proceedings of 8th Conference of the East and Southeast Asian Federation of Soil Science*, Tsukuba, Japan, pp. 8–13.

Humphreys, E., Lewin, L. G., Khan, S., Beecher, H. G., Lacy, J. M., Thompson, J. A., Batten, G. D., Brown, A., Russell, C. A., Christen, E. W. and Dunn, B. W., 2006. Integration of approaches to increasing water use efficiency in rice-based systems in southeast Australia. *Field Crops Research* 97(1), 19–33.

Humphreys, E., Kukal, S. S., Christen, E. W., Hira, G. S. and Sharma, R. K., 2010. 5 Halting the Groundwater Decline in North-West India – Which Crop Technologies will be Winners? *Advances in Agronomy* 109(5), 155–217.

Humphreys, E., Muirhead, W. A., Melhuish, F. M., White, R. J. G. and Blackwell, J. B., 1989. The growth and nitrogen economy of rice under sprinkler and flood irrigation in southeast Australia. II. Soil moisture and mineral N transformations. *Irrigation Science* 10, 201–13.

Husain, A. M., Chowhan, G., Barua, P., Uddin, A. F. M. and Rahman, A. B. M, 2004. Final evaluation report on verification and refinement of the system of rice intensification (SRI) project in selected areas of Bangladesh. PETRRA-Project, IRRI, Dhaka, Bangladesh.

IPCC, 2014. Climate Change 2014: Synthesis Report. Contribution of Working Groups I, II and III to the Fifth Assessment Report of the Intergovernmental Panel on Climate Change [Core Writing Team, R. K. Pachauri and L. A. Meyer (eds.)]. IPCC, Geneva, Switzerland, p. 151.

Jabran, K. and Chauhan, B. S., 2015. Weed management in aerobic rice systems. *Crop Protection* 78, 151–63.

Kadiyala, M. D. M., Mylavarapu, R. S., Li, Y. C., Reddy, G. B. and Reddy, M. D., 2012. Impact of aerobic rice cultivation on growth, yield, and water productivity of rice–maize rotation in semiarid tropics. *Agronomy Journal* 104(6), 1757–65.

Kahlown, M. A., Raoof, A., Zubair, M. and Kemper, W. D., 2007. Water use efficiency and economic feasibility of growing rice and wheat with sprinkler irrigation in the Indus Basin of Pakistan. *Agricultural Water Management* 87(3), 292–8.

Kato, Y. and Katsura, K., 2014. Rice adaptation to aerobic soils: physiological considerations and implications for agronomy. *Plant Production Science* 17(1), 1–12.

Keating, B. A., Carberry, P. S., Bindraban, P. S., Asseng, S., Meinke, H. and Dixon, J., 2010. Eco-efficient Agriculture: Concepts, Challenges, and Opportunities. *Crop Science* 50 (Suppl. 1), S109–S119.

Khush, G. S., 1995. Modern varieties – their real contribution to food supply and equity. *Geojournal* 35(3), 275–84.

Kraus, D., Weller, S., Klatt, S., Santa Bárbara, I., Haas, E., Wassmann, R., Werner, C., Kiese, R. and Butterbach-Bahl, K., 2016. How well can we assess impacts of agricultural land management changes on the total greenhouse gas balance (CO_2, CH_4 and N_2O) of tropical rice-cropping systems with a biogeochemical model? *Agriculture, Ecosystems & Environment*, 224, 104–15.

Kreye, C., Bouman, B. A. M., Castañeda, A. R., Lampayan, R. M., Faronilo, J. E., Lactaoen, A. T. and Fernandez, L., 2009. Possible causes of yield failure in tropical aerobic rice. *Field Crops Research* 111(3), 197–206.

Kumar, V. and Ladha, J. K., 2011. Direct Seeding of Rice: Recent Developments and Future Research Needs. *Advances in Agronomy* 111, 297–413.

Lampayan, R. M., Rejesus, R. M., Singleton, G. R. and Bouman, B. A., 2015. Adoption and economics of alternate wetting and drying water management for irrigated lowland rice. *Field Crops Research* 170, 95–108.

Li, C. S., Qiu, J. J., Frolking, S., Xiao, X. M., Salas, W., Moore, B., Boles, S., Huang, Y. and Sass, R., 2002. Reduced methane emissions from large-scale changes in water management of China's rice paddies during 1980-2000. *Geophysical Research Letters* 29(20), 33–4.

Liang, K., Zhong, X., Huang, N., Lampayan, R. M., Pan, J., Tian, K. and Liu, Y., 2016. Grain yield, water productivity and CH 4 emission of irrigated rice in response to water management in south China. *Agricultural Water Management* 163, 319–31.

Lilley, J. M. and Fukai, S., 1994. Effect of timing and severity of water deficit on four diverse rice cultivars I. Rooting pattern and soil water extraction. *Field Crops Research* 37(3), 205–13.

Linquist, B. A., Anders, M. M., Adviento-Borbe, M. A. A., Chaney, R. L., Nalley, L. L., Da Rosa, E. F. and Kessel, C., 2015. Reducing greenhouse gas emissions, water use, and grain arsenic levels in rice systems. *Global Change Biology*, 21(1), 407–17.

Loomis, R. S. and Connor, D. J., 1992. *Crop Ecology: Productivity And Management In Agricultural Systems*. Cambridge, Cambridge University Press.

Ly, P., Jensen, L. S., Bruun, T. B., Rutz, D. and de Neergaard, A., 2012. The System of Rice Intensification: Adapted practices, reported outcomes and their relevance in Cambodia. *Agricultural Systems* 113, 16–27.

Macadam, R., Drinan, J. and Inall, N., 2002. Building Capacity for Change in the Rice Industry. Rural Industries Research and Development Corporation (RIRDC) Publication No. 02/009. Rural Industries Research and Development Corporation, Canberra, ACT, Australia.

McDonald, D. J., 1994. Temperate rice technology for the 21st century: an Australian example. *Australian Journal of Experimental Agricullture* 34, 877–88.

McDonald, A. J., Hobbs, P. R. and Riha, S. J., 2006. Does the system of rice intensification outperform conventional best management?: A synopsis of the empirical record. *Field Crops Research* 96(1), 31–6.

McIntyre, S., McGinness, H. M., Gaydon, D. and Arthur, A. D., 2011. Introducing irrigation efficiencies: prospects for flood-dependent biodiversity in a rice agro-ecosystem. *Environmental Conservation* 38(03), 353–65.

Molden, D., Murray-Rust, H., Sakthivadivel, R. and Makin, I., 2003. A water productivity framework for understanding and action. In Kijne, J. W., Baker, R. and Molden, D. (Eds), *Water Productivity In Agriculture: Limits And Opportunities For Improvement*. Wallingford UK (CABI), pp. 1–18.

Muirhead, W. A., Blackwell, J. B., Humphreys, E. and White, R. J. G., 1989. The growth and nitrogen economy of rice under sprinkler and flood irrigation in southeast Australia. I. Crop response and N uptake. *Irrigation Science* 10, 183–99.

Mushtaq, S., 2016. Economic and policy implications of relocation of agricultural production systems under changing climate: Example of Australian rice industry. *Land Use Policy* 52, 277–86.

Nie, L., Peng, S., Chen, M., Shah, F., Huang, J., Cui, K. and Xiang, J., 2012. Aerobic rice for water-saving agriculture. A review. *Agronomy for Sustainable Development* 32(2), 411–18.

Palis, F. G., Lampayan, R. M., Bouman, B. A. M., et al., 2014. Adoption and dissemination of alternate wetting and drying technology for boro rice cultivation in Bangladesh. In Kumar, A. (Ed.), *Mitigating Water-ShortageChallenges in Rice Cultiva-tion: Aerobic and Alternate Wetting and Drying RiceWater- Saving Technologies*. IRRI and Asian Development Bank, Manila.

Parthasarathi, T., Mohandass, S., Senthilvel, S. and Vered, E., 2013. Effect of various micro irrigation treatments on growth and yield response of aerobic rice. *International Agricultural Engineering Journal*, 22(4), 49.

Paydar, Z., Gaydon, D. S. and Chen, Y., 2009. A methodology for up-scaling irrigation losses. *Irrigation Science* 27 (5), 347–56.

Peng, S., Bouman, B. A. M., Visperas, R. M., Castañeda, A., Nie, L. and Park, H. K., 2006. Comparison between aerobic and flooded rice in the tropics: Agronomic performance in an eight-season experiment. *Field Crops Research* 96, 252–9.

Peng, S., Buresh, R. J., Huang, J., Zhong, X., Zou, Y., Yang, J., Wang, G., Liu, Y., Hu, R., Tang, Q. and Cui, K., 2010. Improving nitrogen fertilization in rice by site-specific N management. A review. *Agronomy for Sustainable Development* 30(3), 649–56.

Perry, C., Steduto, P., Allen, R. G. and Burt, C. M., 2009. Increasing productivity in irrigated agriculture: agronomic constraints and hydrological realities. *Agricultural Water Management* 96(11), 1517–24.

Qin, J., Hu, F., Zhang, B., Wei, Z. and Li, H., 2006. Role of straw mulching in non-continuously flooded rice cultivation. *Agricultural Water Management* 83(3), 252–60.

Rao, A. N., Johnson, D. E., Sivaprasad, B., Ladha, J. K. and Mortimer, A. M., 2007. Weed management in direct-seeded rice. *Advances in Agronomy* 93, 153–255.

Rijsberman, F. R., 2006. Water scarcity: fact or fiction? *Agricultural Water Management* 80(1), 5–22.

Roberts, L. C., Hug, S. J., Voegelin, A., Dittmar, J., Kretzschmar, R., Wehrli, B., Saha, G. C., Badruzzaman, A. B. M. and Ali, M. A., 2010. Arsenic dynamics in porewater of an intermittently irrigated paddy field in Bangladesh. *Environmental Science & Technology* 45(3), 971–6.

Robertson, G. P., Paul, E. A. and Harwood, R. R., 2000. Greenhouse gases in intensive agriculture: contributions of individual gases to the radiative forcing of the atmosphere. *Science* 289(5486), 1922–5.

Sandhu, B. S., Khera, K. L., Prihar, S. S. and Singh, B., 1980. Irrigation needs and yield of rice on a sandy-loam soil as affected by continuous and intermittent submergence. *The Indian Journal of Agricultural Science* 50 (6), 492–6.

Sheehy, J. E., Peng, S., Dobermann, A., Mitchell, P. L., Ferrer, A., Yang, J., Zou, Y., Zhong, X. and Huang, J., 2004. Fantastic yields in the system of rice intensification: fact or fallacy?. *Field Crops Research* 88(1), 1–8.

Sinha, S. K. and Talati, J., 2007. Productivity impacts of the system of rice intensification (SRI): A case study in West Bengal, India. *Agricultural Water Management* 87(1), 55–60.

Stoop, W. A., Uphoff, N. and Kassam, A., 2002. A review of agricultural research issues raised by the system of rice intensification (SRI) from Madagascar: opportunities for improving farming systems for resource-poor farmers. *Agricultural Systems* 71, 249–74.

Sudhir-Yadav, Gill, G., Humphreys, E., Kukal, S. and Walia, U., 2011a. Effect of water management on dry seeded and puddled transplanted rice. Part 1: Crop performance. *Field Crops Research* 120(1), 112–22.

Sudhir-Yadav, Humphreys, E., Kukal, S., Gill, G. and Rangarajan, R., 2011b. Effect of water management on dry seeded and puddled transplanted rice Part 2: Water balance and water productivity. *Field Crops Research* 120(1), 123–32.

Tabbal, D. F., Bouman, B. A. M., Bhuiyan, S. I., Sibayan, E. B. and Sattar, M. A., 2002. On-farm strategies for reducing water input in irrigated rice; case studies in the Philippines. *Agricultural Water Management* 56(2), 93–112.

Tan, X., Shao, D., Liu, H., Yang, F., Xiao, C. and Yang, H., 2013. Effects of alternate wetting and drying irrigation on percolation and nitrogen leaching in paddy fields. *Paddy and Water Environment* 11(1–4), pp. 381–95.

Thakur, A. K., Uphoff, N. T. and Stoop, W. A., 2015. Scientific Underpinnings of the System of Rice Intensification (SRI): What Is Known So Far? *Advances in Agronomy* 135, 147–79.

Thompson, J. A. and Griffin, D., 2006. Delayed flooding of rice – effect on yields and water. *IREC Farmers Newsletter* 173 (Spring), 52–3.

Thompson, J. P., Mackenzie, J. and Sheedy, G. H., 2012. Root-lesion nematode (Pratylenchus thornei) reduces nutrient response, biomass and yield of wheat in sorghum–fallow–wheat cropping systems in a subtropical environment. *Field Crops Research* 137, 126–40.

Tuong, P., Bouman, B. A. M. and Mortimer, M., 2005. More rice, less water – integrated approaches for increasing water productivity in irrigated rice-based systems in Asia. *Plant Production Science* 8(3), 231–41.

Uphoff, N., Fernandes, E. C. M., Yuan, L. P., Peng, J. M., Rafaralahy, S. and Rabenandrasana, J., 2002. Assessment of the system for rice intensification (SRI). In *Proceedings of the International Conference*, Sanya, China, 1–4 April 2002. Cornell International Institute for Food, Agriculture and Development (CIIFAD), Ithaca, NY.

Wang, H., Bouman, B. A. M., Zhao, D., Wang, C. and Moya, P. F., 2002. Aerobic rice in northern China: opportunities and challenges. In Bouman, B. A. M., Hengsdijk, H., Hardy, B., Bindraban, P. S., Tuong, T. P. and Ladha, J. K. (Eds), *Water-wise Rice Production. Proc International Workshop on Water-wise Rice Production*, 8–11 April 2002, Los Baños, Philippines. International Rice Research Institute, Los Baños, Philippines, pp. 143–54.

Wassmann, R., Neue, H. U., Lantin, R. S., Makarim, K., Chareonsilp, N., Buendia, L. V. and Rennenberg, H., 2000. Characterization of methane emissions from rice fieldsin Asia: II. Differences among irrigation, rainfed, and deepwater rice. *Nutrient Cycling in Agroecosystems* 58, 13–22.

Williams, R. L. and Angus, J. F., 1994. Deep floodwater protects high nitrogen rice crops from low temperature damage. *Australian Journal of Experimental Agriculture* 34, 927–32.

Witt, C., Cassman, K. G., Olk, D. C., Biker, U., Liboon, S. P., Samson, M. I. and Ottow, J. C. G., 2000. Crop rotation and residue management effects on carbon sequestration, nitrogen cycling and productivity of irrigated rice systems. *Plant and Soil* 225, 263–78.

Won, J. G., Choi, J. S., Lee, S. P., Son, S. H. and Chung, S. O., 2005. Water saving by shallow intermittent irrigation and growth of rice. *Plant Production Science* 8, 487–92.

Yang, J., Liu, K., Wang, Z. and Zhu, Q., 2007. Water-saving and high-yielding irrigation for lowland rice by controlling limiting values of soil water potential. *Journal of Integrative Plant Biology* 49, 1445–54.

Yang, S., Peng, S., Xu, J., Luo, Y. and Li, D., 2012. Methane and nitrous oxide emissions from paddy field as affected by water-saving irrigation. *Physics and Chemistry of the Earth, Parts A/B/C* 53, 30–7.

Yao, F., Huang, J., Cui, K., Nie, L., Xiang, J., Liu, X., Wu, W., Chen, M. and Peng, S., 2012. Agronomic performance of high-yielding rice variety grown under alternate wetting and drying irrigation. *Field Crops Research* 126, 16–22.

Yao, Y., Miao, Y., Huang, S., Gao, L., Ma, X., Zhao, G., Jiang, R., Chen, X., Zhang, F., Yu, K. and Gnyp, M. L., 2012. Active canopy sensor-based precision N management strategy for rice. *Agronomy For Sustainable Development* 32(4), 925–33.

Zhang, Q., 2007. Strategies for developing green super rice. *Proceedings of the National Academy of Sciences* 104(42), 16402–9.

Zhang, Z., Zhang, S., Yang, J. and Zhang, J., 2008. Yield, grain quality and water use efficiency of rice under non-flooded mulching cultivation. *Field Crops Research* 108(1), 71–81.

Chapter 5

Improving water management in sorghum cultivation

Jourdan Bell, Texas A&M AgriLife Research and Extension Center, USA; Robert C. Schwartz, USDA-ARS Conservation and Production Research Laboratory, USA; Kevin McInnes, Texas A&M University, USA; Qingwu Xue and Dana Porter, Texas A&M AgriLife Research and Extension Center, USA

1 Introduction

2 Dryland production

3 Irrigation

4 Deficit irrigation

5 Soils and irrigation management

6 Conclusion

7 Where to look for further information

8 References

1 Introduction

Grain sorghum (*Sorghum bicolor* [L.] Moench) is a drought-tolerant crop that has been grown across the American Great Plains since the early twentieth century, primarily as a feed crop. Physiologically, grain sorghum is well adapted to the semi-arid conditions of the American Great Plains because it can withstand periods of water stress. Under ideal conditions and management practices, modern hybrids can yield over 11 000 kg ha^{-1}, but in semi-arid regions, plant available water is a major factor limiting crop production (Abunyewa et al., 2011). While early cultivation of grain sorghum was under dryland production, it was recognized that the efficient use of water, including precipitation and stored soil moisture, was critical for improved performance. Evaluation of sorghum water use efficiencies (WUEs; biomass production including grain yield per unit of water used) was reported as early as 1914 by Briggs and Shantz (1914), who noted improved WUEs of sorghum in comparison to other grain crops. Numerous authors have since defined the improvements in WUEs as a result of sorghum being a C4 plant with a high transpiration efficiency (Xin, 2009; Rooney, 2004; Mortlock and Hammer, 1999; Boyer, 1996). As a C4 plant, sorghum produces more biomass per unit of water transpired than cool-season or C3 plants because C4 plants have a more efficient photosynthetic rate. This enables C4 plants such as sorghum to lose less CO_2 during photorespiration, resulting in greater conversion of CO_2 to biomass. In addition to being a C4 plant, sorghum is more drought tolerant than other C4 plants such as corn due to a deep root system and epicuticular wax to minimize transpirational water losses. Sorghum's deep, fibrous root system provides greater root volume and therefore potential access to a greater soil water volume, enabling sorghum to withstand greater periods of water stress if there is ample stored soil water. Additionally, sorghum is able to maintain turgor pressure in cells through the regulated accumulation of solutes (osmotic adjustment) (Ferres et al., 1978; Assefa, 2010). Under water deficits, the low osmotic potential in the roots reduces the water movement through the plant, resulting in less water loss through the stomatal openings (Fekade and Daniel, 1992). Furthermore, Sanchez-Diaz and Kramer (1971) concluded that after periods of water stress, the plant water potential and saturation potential recovered more quickly than corn. While sorghum is physiologically adapted to withstand prolonged periods of water stress, such tolerances can come at the expense of reduced yield (Assefa, 2010; Peacock, 1982; Fereres et al., 1978).

With the expansion of the fed beef cattle industry on the Southern High Plains in the 1970s as well as corn's greater yield potential and WUE under full irrigation, corn replaced sorghum on much of the US irrigated acreage. However, grain sorghum is still the predominant summer grain crop under dryland and limited irrigation. In the last 25 years, US sorghum production has ranged from 1733 to 7160 MT, Mg (214–884 million bushels) with 75% of the US sorghum production occurring in Kansas and Texas (USDA-NASS, 2017). Consequently, the two lowest production years (2011 and 2012) coincided with record droughts in Kansas and Texas (Fig. 1); during this period, annual precipitation ranged from 102 to 260 mm. Regionally, mean annual precipitation across the US sorghum-producing regions (Fig. 2) ranges from 380 to 800 mm from the Texas Southern High Plains to central Kansas. To minimize yield losses accrued during periods of water stress for optimal production, adoption of improved strategies under both dryland and irrigated systems is essential.

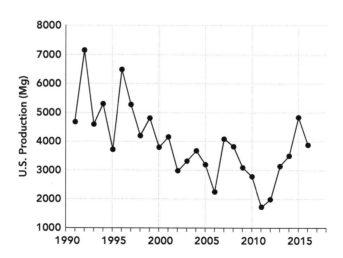

Figure 1 25 years of US grain sorghum production (USDA-NASS, 2017).

Improving water management in sorghum cultivation

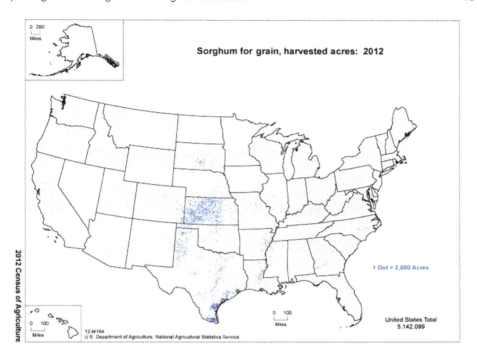

Figure 2 US 2012 grain sorghum harvested acres (USDA Census of Agriculture, 2012).

2 Dryland production

Historical grain sorghum yield data at the USDA-ARS Conservation and Production Research Laboratory in Bushland revealed that sorghum yields increased 139% between 1939 and 1997 (Unger and Baumhardt, 1999). While the authors recognized the importance of hybrid improvement in the yield increase, 67% of the increase was attributed to improved soil water content at planting because of reduced tillage and residue management.

In dryland production systems, conservation tillage and residue management have been essential to yield stabilization by increasing stored soil water. On the Great Plains, the wheat–fallow–sorghum rotation is widely adopted under both dryland and limited irrigation. This rotation sequence balances production and the fallow period efficiency to increase stored soil water at planting. By definition, conservation tillage leaves 30% of the crop residue on the soil surface after tillage and after the subsequent crop is planted (Unger and Baumhardt, 1999). Improvements in residue management were initially in response to widespread soil erosion resulting from the drought of the 1930s and the Great American Dust Bowl, but scientists and farmers quickly recognized that maintaining soil residue also minimized evaporative (E) losses from the soil surface and increased soil water contents at planting. Conservation tillage practices commonly used in dryland sorghum production are no-till (NT) and stubble-mulch tillage (SMT). Because weeds utilize valuable soil water, weed control is critical to the success of the conservation tillage practices (Wise and Army, 1960), especially in NT dryland systems. NT fully relies on chemical herbicides for weed control because tillage is eliminated. Unger and Baumhardt (1999) reported that NT is often considered the 'ultimate' conservation tillage method, but they recognize that other conservation tillage methods such as SMT leave sufficient residue on the soil surface to minimize soil erosion and evaporation. Previous NT research has demonstrated that yields are often comparable to yields with other conservation practices on fine-textured soils in the semi-arid Great Plains of Texas (Baumhardt et al., 2017; Jones et al., 1994), but in dry years with low biomass production, yields are reduced under NT. This is often pronounced in semi-arid dryland systems with fine-textured soils due to soil crust formation and surface sealing (Baumhardt et al., 2017; Jones et al., 1994). Schwartz et al. (2003) reported that crust formation decreased infiltration and soil water storage in these soils. With SMT, V-shaped blades cut weed roots approximately 10 cm below the soil surface without inversion and therefore retain residue on the soil surface. In a long-term research study conducted by Baumhardt et al. (2017) from 1983 to 2013, the annualized cumulative run-off during the fallow period averaged 27.4 mm for NT and 17.3 mm for SMT; however, evaporative losses were lower under NT, resulting in increased precipitation storage.

Planting geometry has been evaluated as a method to improve crop water use and mitigate stress. Early research indicated that grain sorghum planted in narrow rows produced greater biomass and grain yields in humid and subhumid climates as well as under full irrigation, but in semi-arid regions or in years with water stress, there was a greater risk for yield reduction as a result of increased plant competition for spatially variable stored soil water (Fernandez et al., 2012; Jones and Johnson, 1991; Staggenborg et al., 1999; Steiner, 1986; Grimes and Musick, 1960).

3 Irrigation

The expansion of internal combustion engines and turbine pumps led to widespread adoption of irrigation that stabilized and increased production in the semi-arid Great Plains (Colaizzi et al., 2009; Howell, 2006; Rhoades, 1997; Musick et al., 1988). Grain sorghum acreage peaked during the 1950s as a result of extensive irrigation, but yields continued to increase with advances in agronomic and irrigation management in addition to improved grain sorghum hybrids. Improvements in irrigation management including the transition from gravity-fed irrigation with graded furrows to sprinkler irrigation, and improved application efficiencies, further increased sorghum yield per acre as well as yield per unit of water applied (irrigation water use efficiency, IWUE). While improvements in irrigation technology have greatly increased IWUE, advancements in irrigation management strategies will likely provide the most extensive water savings (Garces-Restrepo et al., 2007). The lack of precipitation can potentially magnify yield responses to irrigation treatments (Allen and Musick, 1990; Howell, 2007). Under water-limited conditions, improvements in IWUE coupled with practices to improve precipitation and soil water management are necessary to optimize the crop WUEs (crop production per unit of water consumed). Crop WUE accounts for crop utilization of stored soil water, precipitation and irrigation.

In much of the US Great Plains, irrigation water is supplied by the Ogallala Aquifer (Fig. 3). The Ogallala underlies portions of eight states, with the saturated thickness currently in

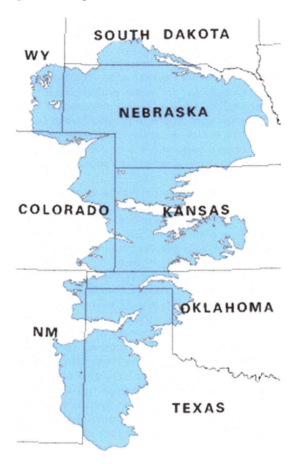

Figure 3 Distribution of the Ogallala Aquifer in the Great Plains (https://en.wikipedia.org/wiki/File:Ogallala_Aquifer_map.png).

decline from Kansas south through Texas because of significant withdrawals and negligible recharge rates. Consequently, irrigation water from the Ogallala Aquifer is becoming less reliable due to the subsequent decline in well capacities in addition to increased competition for water supplies between urban and rural sectors. Traditionally, irrigation has been applied to fully meet the seasonal crop water demand; however, future water use will be limited either by restricted pumping or by reduced well capacities. To sustain domestic and industrial requirements, agricultural water use is projected to decrease by up to 34% by 2060 in the Texas High Plains due to ensuing policies that will restrict pumping (Wagner, 2012). Utilization of dryland management strategies with limited irrigation rather than irrigating to meet the full crop water demand could potentially stabilize crop production and support an agriculturally driven economy for an extended period (Bordovsky et al., 2011).

Combining conservation tillage practices and improved irrigation management strategies often allows producers to maximize the soil water storage from variable precipitation to improve WUE under conditions of limited irrigation water.

With limited water for irrigation, deficit irrigation (DI) is the most common irrigation strategy used for sorghum production. Howell et al. (2007) demonstrated that DI grain sorghum established on a nearly full soil water profile permitted enhanced soil water extraction and only marginal declines in yield compared to fully irrigated sorghum. However, frequent and shallow applications of irrigation water (e.g. ≤20 mm) during the growing season can result in shallow sorghum root development (Myers et al., 1984), which may reduce water uptake deeper in the profile. Moreover, both the timing and amount of irrigation during the growing season are critical to the success of DI strategies (Bell, 2014).

Irrigation scheduling often utilizes external estimates of crop evapotranspiration (ETc) and/or measurements of soil water depletion to estimate crop water requirements, while ETc equates the crop water use as a combination of evaporation (E) from the soil surface and transpiration (T) for plant growth. ETc is calculated using an appropriate crop-specific coefficient (Kc) and the reference evapotranspiration (ET_o). ET_o is the calculated evapotranspiration using weather station data including air temperature, relative humidity, precipitation, wind speed and solar irradiance. Oftentimes, local weather data are unavailable and therefore ET-based scheduling is not practical, so ET can be calculated directly through measurements in the changes in water content within the root zone. In practice, however, soil water sensors are not normally installed throughout the rooting zone and often are of insufficient accuracy to estimate ETc and hence irrigation is scheduled based on the attainment of established soil water depletion levels or water potentials at one or more depths in the most active regions of the root zone. Advancements in electronic sensing of soil moisture provides irrigators information they can use to assess the level of soil water depletion or increase and to determine when to initiate irrigation as well as how much water to apply to avoid water stress and minimize drainage below the root zone. Irrigation can be scheduled based on ETc and soil water monitoring or a combination of the two methods for successful grain sorghum yield (Bell, 2014).

Utilization of ETc is facilitated through on-farm weather stations or regional ET networks that distribute data on both the daily weather and ETc for irrigation scheduling. Daily ETc data permit producers to make daily adjustments based on the changes in ETc rather than irrigating at a constant rate. Wireless technologies and Internet integration enhance the application of ETc for irrigation scheduling by providing producers access to real-time data to improve irrigation efficiency and integrate irrigation with precision technologies (Pierce and Elliott, 2008). Real-time data permit the producer to balance the daily environmental demands with available soil moisture and irrigation capacity to minimize overwatering the crop, thus improving economic returns by reducing irrigation costs. Online ET-based irrigation scheduling tools, downloadable ET calculators and smartphone apps for irrigation scheduling utilize ET_o data from regional ET networks along with producer inputs to determine ETc (Rogers, 2012; Migliaccio et al., 2015). Just a few of the many user inputs are crop type, irrigated area, irrigation system type and efficiency, row spacing, rooting depth, soil type and water conservation mode. Migliaccio et al. (2015) noted that user-friendly apps can provide producers notifications on factors such as the daily water balance and weather forecasts that prompt a response to a critical event.

4 Deficit irrigation

Comprehensive irrigation strategies are necessary to optimize limited water supplies. Irrigation management should supplement seasonal precipitation and stored soil water to a degree sufficient to meet crop requirements while maximizing net returns and minimizing run-off and drainage below the root zone. Irrigation scheduling strives to manage irrigation timing and rates to efficiently satisfy crop water demand such that periods of excessive water stress associated with yield reductions are minimized. Regulated or managed DI enables the irrigator to apply water at a reduced rate throughout selected periods of the growing season in an attempt to optimize the rate of return. DI may increase the WUE of some crops (English, 1990; Fereres and Soriano, 2007; Musick et al., 1994); however, judicious scheduling of irrigation combined with appropriate and well-maintained technology is necessary to supply water at deficit levels during non-critical growth stages while ensuring adequate irrigation during critical growth stages to mitigate water stress (Perry et al., 2009). Vadez et al. (2013) concluded that research results from many crops grown under DI strategies suggest that incorporation of managed deficit irrigation strategies would shift crop water use from the vegetative to the reproductive stage to increase yield under water-limited conditions.

Howell and Hiller (1975) concluded that the yield response of grain sorghum to irrigation was most sensitive to having water available at critical times. Seasonal crop water requirements coincide with the growth patterns (Fig. 4) tempered by the atmospheric water demands during the growing season. Traditionally, DI entailed irrigation throughout the entire growing season targeting a fraction of crop water requirements compared with full irrigation (FI) (Howell, 2007). However, Bell (2014) evaluated a managed deficit irrigation (MDI) strategy whereby a fraction of the full irrigation requirement was applied at critical growth stages. Using MDI, early season irrigation was either reduced or eliminated; but a greater percentage of the daily crop water demand (ETc) was applied during the reproductive stages from growing point differentiation to half-bloom. Although the greatest grain yields and WUEs were achieved with FI, irrigation under MDI reduced WUE in only one of the three years. In contrast, yields and WUEs of DI were reduced in all years compared to FI.

With grain sorghum, it has been well documented (Van Oosterom and Hammer, 2008; Tolk et al., 2013; Prasad et al., 2008; Blum, 1996; Crauford and Peacock, 1993; Peacock, 1982; Eck and Musick, 1979) that water stress extending from growing point differentiation through half-bloom suppresses grain yield due to reduced seeds per panicle. Quantifying the effects of the magnitude of water stress at critical growth stages on the resultant ETc and grain yield is essential to evaluate irrigation strategies and improve WUEs. It is also imperative that under conditions of limited water, precipitation is considered in designing the irrigation management plan.

Growing point differentiation is the initial stage of reproductive development in grain sorghum (Vanderlip, 1979). At growing point differentiation, the crop begins to develop rapidly – 7–10 leaves have expanded and rapid nutrient uptake has begun. Water stress during growing point differentiation can reduce the number of seeds developed per panicle. Water stress during the boot stage minimizes head extension from the flag leaf sheath, which reduces pollination (Gerick et al., 2003). Half-bloom is the final stage of reproductive development where approximately half of the sorghum plants in a field have begun flowering. During half-bloom, water stress may induce floral abortion and decrease

Figure 4 Generalized pattern of grain sorghum that coincides with periods of crop water demand.

grain yield. In contrast, water stress from anthesis through the dough stage may reduce grain/seed mass (Ockerby et al., 2001; Maman et al., 2004).

Irrigation scheduling should also utilize knowledge of the crop rooting depth, soil physical properties, irrigation system capabilities/constraints and irrigation water quality to optimize yield. Knowledge of the crop rooting depth can aid the irrigator in determining the potential volume of plant available water and the allowable depletion when scheduling irrigation. Under dryland, research has shown that the sorghum root

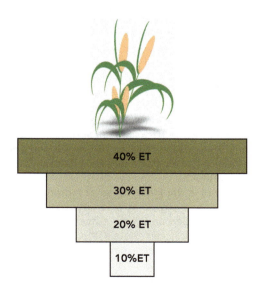

Figure 5 Generalized depiction of soil water extraction by rooting depth.

© Burleigh Dodds Science Publishing Limited, 2018. All rights reserved.

system often reaches depths over six feet (Moroke et al., 2005). However, it is important for producers to recognize that while deep soil moisture is available, research has shown that approximately 40% of water used originates in the upper 25% of the root zone (Fig. 5).

5 Soils and irrigation management

Soil properties such as texture and structure directly affect irrigation. The soil water-holding capacity as well as water flow into and through the soil and the rate and frequency at which irrigation is applied are functions of the soil type. Consequently, knowledge of soil physical properties is essential for successful irrigation scheduling. Soil textural classes are based on the fractions of sand, silt and clay particles. Soil structure is the arrangement of the sand, silt and clay particles into aggregates defined as granular, platy, blocky, columnar and prismatic. Aggregates are formed and their properties are altered through both natural and human activities such as wetting, freezing, thawing, organic matter decomposition and tillage. The aggregate arrangement affects the soil porosity (pore spaces in the soil) and thus water storage and flow in the soil. Coarse-textured soils with a high percentage of sand have large pores, while fine-textured soils with a high percentage of silts and clays have many small pores. Water drains freely through large pores. In comparison, capillary forces retain water in small pores; consequently, the clay fraction is the most important component affecting a soil's water-holding capacity due to its influence on structure and porosity. As the clay fraction increases, the quantity of water held in the soil pores after drainage increases. Soil texture and porosity can be directly measured by the soil bulk density, the weight of soil including solids and pores per unit volume. Soil texture and structure affect the infiltration rate and the permeability of water into the soil. Infiltration is the rate at which water moves into the soil, while permeability is the rate at which water moves through the soil. Fine-textured soils have a low infiltration rate. In comparison, coarse soils have a high infiltration rate. The infiltration rate directly affects the rate of run-off. If the irrigation is applied at a rate greater than the infiltration rate, run-off will occur. In addition to soil texture and structure, infiltration and permeability are also governed by soil organic matter, soil compaction and the antecedent soil moisture. Wet soils have lower infiltration and permeability than dry soil. Infiltration and permeability also affect drainage below the root zone. It is the responsibility of the irrigator to manage irrigation to minimize drainage, run-off and evaporation (E) to increase stored soil water for transpiration (T). The irrigator should have knowledge of the soil so that the irrigation rate does not exceed the infiltration rate (Table 1). DI should be managed according to the soil

Table 1 Average infiltration rate for various soil textural classes

Textural class	Infiltration rate (mm hr^{-1})	Infiltration rate (in hr^{-1})
Clay	2.5–5.1	0.1–0.2
Silt loam	6.4–10.2	0.2–0.4
Sandy loam	8.9–12.7	0.4–0.5
Fine sand	12.7–17.8	0.5–0.7

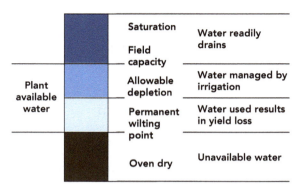

Figure 6 Categories of soil water.

texture. On fine-textured soils, the irrigation efficiency is maintained by applying irrigation at greater depths and time intervals, which minimizes soil evaporation and maximizes soil water storage. Conversely, on sandy soils where drainage is a concern, shallow, frequent irrigations are recommended to avoid drainage below the root zone.

Soil water content is generally expressed as a per cent volume, or even a 'depth of water per depth of soil'. Some important soil water content reference points are saturation, field capacity, plant available water and permanent wilting point. While a soil may have a known amount of water, only a percentage of the water is available to the plant (Fig. 6). The volume of water held by the soil between field capacity and wilting point is the plant available water. Plant available water is the fraction of the stored soil water that can be utilized by plants. This fraction varies by soil textural class due to pore size and the force (suction) with which water is held in the soil (Table 2). While plant available water can be utilized by the crop, irrigators often plan for a portion of the plant available water to be utilized by the crop (managed allowable depletion) in order to minimize plant stress and production losses that occur when plants expend energy, utilizing the lower limits of the plant available water. Sorghum yields are reduced when the allowable depletion is greater than 50% of the plant available water (Bell, 2014; Doorenbos and Pruitt (1977). Permanent wilting point is the lower limit where plants are unable to absorb water. In addition to being directly related to soils, the allowable depletion is a function of the atmospheric conditions that determine the crop water use (Tolk and Howell, 2008; Robertson and Fukai, 1994).

Table 2 Average plant available water for different soil textural classes

Textural class	Plant available water (mm m^{-1})	Plant available water (in ft^{-1})
Clay	101–126	1.2–1.5
Silty clay loam	151–168	1.8–2.0
Fine sandy loam	126–168	1.5–2.0
Fine sand	63–84	0.8–1.0
Coarse sand	21–63	0.3–0.8

6 Conclusion

Efficient water management is dependent on management decisions based on soil type, irrigation system limitations, crop species and production goals. While irrigation enhances production and reduces risk, especially in semi-arid regions or regions with temporally variable precipitation, there are also additional costs associated with irrigation including equipment, labour and energy costs. Future irrigation practices and methods of crop selection must be adopted to prolong groundwater resources. As irrigation water across Texas becomes limited by declining aquifer levels and/or restricted by groundwater district regulations, it is important that crop selection and crop water demands do not exceed the water availability. Because sorghum can withstand periods of drought, it will continue to be a valuable cropping option for prolonging limited water resources.

Globally, increasing opportunities for utilization of sorghums as livestock feed in addition to expanding food and industrial markets presents farmers in water limited regions greater production opportunities. Consequently, there are extensive research opportunities to evaluate agronomic practices to optimize sorghum quality in addition to agronomic yield while increasing crop WUE. Previous research has extensively addressed management of water resources to optimize agronomic yields; however, there are significant differences in grain quality because of drought and water stress. Research to optimize grain quality and properties is an important research area for expanding and securing future markets. Additionally, there are extensive opportunities to evaluate the deficit irrigation strategies and irrigation scheduling using emerging irrigation technologies.

7 Where to look for further information

Additional information about water management and irrigation technologies can be found through the Research and Extension Services of Texas A&M AgriLife, Kansas State University and the University of Nebraska. Specific resources are located at the Texas A&M AgriLife Water Education Network (https://water.tamu.edu/water-management-irrigation/) and Kansas State University Department of Agronomy Grain Sorghum page (https://www.agronomy.k-state.edu/extension/crop-production/grain-sorghum/). Scientific journals such as *Agronomy Journal, Crop Science, Irrigation Science, Agricultural Water Management*, and *Transactions of ASABE Irrigation* are popular journals where one can locate recent research concerning water management.

8 References

Abunyewa, A. A., R. B. Ferguson, C. S. Wortman, D. J. Lyon, S. C. Mason, S. Irmak and R. N. Klein. 2011. Grain sorghum water use with skip-row configuration in the Central Great Plains of the USA. *Afr. J. Agric. Res.* 6(23):5328–38.

Allen, R. R. and J. T. Musick. 1990. Effect of tillage and preplant irrigation on sorghum production. *Appl. Eng. Agric.* 6(5):611–18.

Assefa, Y., S. A. Staggenborg and V. P. V. Prasad. 2010. Grain sorghum water requirement and responses to drought stress: A review. *Crop Manag.* Online. doi:10.1094/CM-2010-1109-01-RV.

Baumhardt, R. L., R. C. Schwartz, O. R. Jones, B. R. Scanlon, R. C. Reedy and G. W. Marek. 2017. Long-term conventional and no-tillage effect on field hydrology and yields of a dryland crop rotation. *Soil Sci. Soc. Am. J.* 81:200–9.

Bell, J. M. 2014. Responses of grain sorghum to profile and temporal dynamics of soil water in a semi-arid environment. Ph.D. Thesis, Texas A&M University, College Station, TX.

Bordovsky, J. P., J. T. Mustian, A. M. Cranmer and C. L. Emerson. 2011. Cotton-grain sorghum rotation under extreme deficit irrigation conditions. *Appl. Eng. Agric.* 27:359–71.

Boyer, J. S. 1996. Advances in drought tolerance in plants. *Adv. Agron.* 56:187–218.

Briggs, L. J. and Shantz, H. L. (1914). Relative water requirement of plants. *J. Agric. Res. (Washington, DC)* 3:1–63.

Childs, S. W. and R. J. Hanks. 1975. Model of soil salinity effects on crop growth. *Utah Agric. Exp. Sta. Journal Paper no. 1920.*

Colaizzi, P. D., P. H. Gowdam, T. H. Marek and D. O. Porter. 2009. Irrigation in the Texas High Plains: A brief history and potential reductions in demand. *Irrig. Drain.* 58(3):257–74.

Crauford, P. Q. and J. M. Peacock. 1993. Effect of heat and drought stress on sorghum (*Sorghum bicolor*). II. Grain yield. *Exp. Agric.* 29:77–86.

Doorenbos, J. and W. H. Pruitt. 1977. Crop water requirements. *FAO Irrigation and Drain Paper 24.* FAO, Rome, Italy, 144pp.

English, M. J. 1990. Deficit irrigation. I: Analytical framework. *J. Irrig. Drain. Eng.* 116(3):399–412.

Fekade, S. G. and R. K. Daniel. 1992 Osmotic adjustment in sorghum. *Plant Physiol.* 99:577–82.

Fereres, E., E. Acevedo, D. W. Henderson and T. C. Hsiao. 1978. Seasonal changes in water potential and turgor maintenance in sorghum and maize under water stress. *Physiol. Plant.* 44:261–7.

Garces-Restrepo, C., D. Vermillion and G. Muoz. (2007). *Irrigation Management Transfer: Worldwide Efforts and Results.* Food and Agriculture Organization of the United Nations, Rome.

Gerik, T., B. Bean and R. Vanderlip. 2003. Sorghum growth and development. Texas Cooperative and Extension Publication B-6137.

Gollehon, N. and B. Winston. 2013. Groundwater irrigation and water withdrawals: The Ogallala Aquifer initiative. Economic series number 1. USDA-NRCS, Washington DC, USA. http://www.nrcs.usda.gov/Internet/FSE_DOCUMENTS/stelprdb1186440.pdf.

Grattan, S. R. Irrigation water salinity and crop production. University of California Publication 8066. FWQP Reference Sheet 9.10.

Grimes, D. W. and J. T. Musick. 1960. Effect of plant spacing, fertility, and irrigation managements on grain sorghum production. *Agron. J.* 52:647–50.

Howell, T. A. 2006. Challenges in increasing water use efficiency in irrigated agriculture. In: *International Symposium on Water and Land Management for Sustainable Irrigated Agriculture*, Adana, Turkey, 4–8 April 2006.

Howell, T. A. and E. A. Hiller. 1975. Optimization of water use efficiency under high frequency irrigation I. Evapotranspiration and yield relationship. *Trans. ASAE.* 18:873–8.

Howell, T. A., J. A. Tolk, S. R. Evett, K. S. Copeland and D. A Dusek. 2007. Evapotranspiration of deficit irrigated sorghum and winter wheat. In: Clemmens, A. J. (Ed.), *USCID Fourth International Conference on Irrigation and Drainage. The Role of Irrigation and Drainage in a Sustainable Future*, 3–6 October 2007, Sacramento, California, pp. 223–39. 2007 CDROM.

Jones, O. R. and G. L. Johnson. 1991. Row width and plant density effects on Texas High Plains sorghum. *J. Prod. Agric.* 4:613–21.

Jones, O. R., V. L. Hauser and T. W. Popham. 1994. No-tillage effects on infiltration, runoff, and water conservation on dryland. *Trans. ASABE* 37(2):473–9.

Maman, N., S. C. Mason, D. J. Lyon and P. Dhungana. 2004. Yield components of pearl millet and grain sorghum across environments in the Central Great Plains. *Crop Sci.* 44:2138–45.

Migliaccio, K. W., K. T. Morgan, G. Vellidis, L. Zotarelli, C. Fraisse, B. A. Zurweller, J. H. Andreis, J. H. Crane and D. L. Rowland. 2015. Smartphone apps for irrigation scheduling. *Trans. ASABE* 59(1):291–301.

Moroke, T. S., R. C. Schwartz, K. W. Brown and A. S. R. Juo. 2005. Soil water depletion and root distribution of three dryland crops. *Soil Sci. Soc. Am. J.* 69:197–205.

Mortlock, M. Y. and G. L. Hammer. 1999. Genotype and water limitation effects on the transpiration efficiency in sorghum. *J. Crop Prod.* 2:265–86.

Musick, J. T., F. B. Pringle and J. D. Walker. 1988. Sprinkler and furrow irrigation trends—Texas High Plains. *Appl. Eng. Agric.* 4(1):46–52.

Norwood, C. A., A. J. Schlegel, D. W. Morishita and R. E. Gwin. 1990. Cropping system and tillage effects on available soil water and yield of grain sorghum and winter wheat. *J. Prod. Agric.* 3:356–62.

Ockerby, W. E., D. J. Midmore and D. F. Yule. 2001. Leaf modification delays panicle initiation and anthesis in grain sorghum. *Aust. J. Agric. Res.* 52:127–35.

Peacock, J. M. 1982. Response and tolerance of sorghum to temperature stress. In: *Sorghum in the Eighties: Proceedings of the International Symposium on Sorghum*, 2–7 November 1981, Patancheru, Andhra Pradesh, India.

Perry, C., P. Steduto, R. G. Allen and C. M. Burt. 2009. Increasing productivity in irrigated agriculture: Agronomic constraints and hydrological realities. *Agric. Water Manag.* 96(11):1517–24.

Pierce, F. J. and T. V. Elliott. 2008. Regional and on-farm wireless sensor networks for agricultural systems in Eastern Washington. *Comput. Electron. Agric.* 61(1):32–43.

Prasad, P. V., S. R. Pisipati, R. N. Mutava and M. R. Tuinstra. 2008. Sensitivity of grain sorghum to high temperature stress during reproductive development. *Crop Sci.* 48:1911–17.

Rhoades, J. D. 1997. Sustainability of irrigation: An overview of salinity problems and control strategies, pp. 1–42. In: *CWRA 1997 Annual Conf. 'Footprints of Humanity: Reflections on Fifty Years of Water Resource Developments'*, Lethbridge, Alberta, Canada, 3–6 June 1997.

Robertson, M. J. and S. Fukai. 1994. Comparison of water extraction models for grain sorghum under continuous soil drying. *Field Crops Res.* 36:145–60.

Rogers, D. H. 2012. Introducing the web-based version of KanSched: An ET-based irrigation scheduling tool. In: *Proceedings of the 24th Annual Central Plains Irrigation Conference*, Colby, Kansas, 21–22 February 2012.

Rooney, W. L. 2004. Sorghum improvement-integrating traditional and new technology to produce improved genotypes. *Adv. Agron.* 83:37–109.

Sanchez-Daiz, M. and P. J. Kramer. 1971. Behavior of com and sorghum under water stress and during recovery. *Plant Physiol.* 48:613–16.

Schwartz, R. C., S. R. Evett and P. W. Unger. 2003. Soil hydraulic properties of cropland compared with reestablished and native grassland. *Geoderma* 116:47–60.

Staggenborg, S., B. Gordon, R. Taylor, S. Duncan and D. Fjell. 1999. Narrow-row grain sorghum production in Kansas. Kansas State University Agricultural Experiment Station and Cooperative Extension Service Publication MF-2388.

Steiner, J. L. 1986. Dryland grain sorghum water use, light interception, and growth responses to planting geometry. *Agron. J.* 78:720–6.

Tolk, J. A. and T. A. Howell. 2008. Field water supply: Yield relationships of grain sorghum grown in the USA Southern Great Plains soils. *Agric. Water Manag.* 95(12):1303–13.

Tolk, J. A., T. A. Howell and F. R. Miller. 2013. Yield component analysis of grain sorghum grown under water stress. *Field Crops Res.* 145:44–51.

Unger, P. W. and R. L. Baumhardt. 1999. Factors related to dryland grain sorghum yield increases: 1939 through 1997. *Agron. J.* 91:870–5.

Unger, P. W. and R. L. Baumhardt. 2001. Historical development of conservation tillage in the southern Great Plains. In: Stiegler, J. H. (Ed.), *Proceedings of the 24th Annual Southern Conservation Tillage Conference for Sustainable Agriculture*, Oklahoma City, July 2001. Oklahoma State University, Oklahoma City, OK.

USDA. 2012 *Census of Agriculture. Summary and State Data: Volume 1: Geographic Area Series*. National Agricultural Statistics Service, Washington DC.

USDA-NASS. 2017. Crop production: 2016 summary. ISSN: 1057-7823. http://usda.mannlib.cornell.edu/usda/current/CropProdSu/CropProdSu-01-12-2017.pdf.

Vadez, V., J. Kholova, R. S. Yadav and C. T. Hash. 2013. Small temporal differences in water uptake among varieties of pearl millet (*Pennisetum glaucum* (L.) R. Br.) are critical for grain yield under terminal drought. *Plant Soil* 371:447–62.

Vanderlip, R. L. 1979. How a sorghum plant develops (No. C040. 065). Cooperative Extension Service, Kansas State University.

Wagner, K. 2012. Status and trends of irrigated agriculture in Texas: A special report. Texas Water Resources Institute, College Station, TX.

Xin, Z., R. Aiken and J. Burke. 2009. Genetic diversity of transpiration efficiency in sorghum. *Field Crops Res.* 111:74–80

Ingram Content Group UK Ltd.
Milton Keynes UK
UKHW022044210723
425584UK00010B/47